Technical Writing

What It Is & How To Do It

Julie M. Zeleznik
Iowa State University

With Contributions from

Philippa J. Benson
National Institutes of Health

Rebecca E. Burnett
Iowa State University

NEW YORK

 Published in the United States by LearningExpress, LLC, New York.

Library of Congress Cataloging-in-Publication Data

Zeleznik, Julie M.

Technical writing : what it is and how to do it / by Julie M. Zeleznik

p. cm.

Includes bibliographical references (p.).

ISBN 1-57685-267-9

1. Technical writing. I. Title.

T11.Z45 1999

808'.0666—dc21 99-38274

CIP

Printed in the United States of America

9 8 7 6 5 4 3 2 1

First Edition

For Further Information

For information on LearningExpress, other LearningExpress products, or bulk sales, please write to us at:

LearningExpress®
900 Broadway
Suite 604
New York, NY 10003

Visit LearningExpress on the World Wide Web at www.LearnX.com

ISBN 1-57685-267-9

Contents

About the Authors

Julie M. Zeleznik, M.A., is a doctoral student in the Rhetoric and Professional Communication program at Iowa State University. She has taught courses in technical commnication, business communication, and college composition.

Philippa Benson, Ph.D., is Information Officer/Technical Editor for the National Institutes of Health. She has taught and consulted in technical and scientific writing and writing across the disciplines at Georgetown University, Carnegie Mellon University, and the University of Pittsburgh.

Rebecca Burnett, Ph.D., is Professor of Rhetoric and Professional Communication in the Department of English at Iowa State University. She has published numerous articles in scholarly journals. She is editor of the *Journal of Business and Technical Communication* and author of a widely used textbook, *Technical Communication.*

Part I

Understanding the Nature of Technical Communication

Chapter 1

What Is Technical Communication?

To understand the importance of technical communication, you must know how to define and characterize it. Identifying what technical communication produces, who produces it, and who benefits helps you to discover if technical communication is the career field for you.

The communication skills that technical writers use every day—whether these skills involve writing accurate and complete documents, designing usable and appealing Web sites, or creating organized and interesting oral presentations—are more highly in demand now than ever before.

The 1998 "20 Hot Job Tracks" issue of *U.S. News and World Report* included as one of its articles on employment opportunities for 1999 an article about the diversity and growth of the field of technical communication. The article cited that one professional organization for technical writers, the Society for Technical Communication, saw "its membership rise 53 percent since 1990, from 13,159 to 20,190, as the demand for technology filters down from the lab to the home" (85).

This book introduces you—the new high school graduate, the student returning to college, or the professional seeking a career change—to the rapidly growing field of technical communication. In this chapter, we define technical communication by discussing who performs it, what it produces, and who benefits from it. After you have read this chapter, you will know the answers to these basic and important questions about technical communication:

- What is *technical communication?*
- Where is technical communication *performed?*

- What are the *products* of technical communication?
- What are the *characteristics* of technical communication?
- Who *benefits* from effective technical communication?

Discovering the answers to these questions gives you the necessary background to learn more about this field. In the subsequent chapters, you will learn about how technical writers collaborate with business colleagues, how they use technology to communicate, and how they plan, draft, revise, and edit documents.

What is technical communication?

Technical communication is the process by which professionals in business communicate with one another, their customers, and their clients. In other words, technical communication enables professionals to do the work that makes their businesses run.

All business professionals, not just technical writers, communicate all of the time—through written means, such as print or on-line documents, and through oral means, such as presentations, meetings, and conversations. These means of communication enable professionals to implement, direct, change, move forward with, and complete their work.

For example, a team of engineers makes steady progress on the design of an improved automobile braking system. To report this progress to their supervisors and fellow engineers (who are designing other components for the same automobile), the team writes bi-weekly progress reports describing the work they have completed and the work they need to perform. This form of technical communication helps the team communicate with their colleagues and helps their colleagues understand the team's progress.

Professionally trained men and women who facilitate communication in the workplace are called *technical writers*. This title may be misleading, however, as their jobs always involve more than simply writing documents—their jobs are about communication. Technical writers are trained to help business professionals communicate with one another, their clients, and customers.

Where is technical communication performed?

Since every occupation relies on communication, professional technical writers work nearly everywhere—from business- and finance-related fields to the engineering, manufacturing, and publishing sectors. Technical writers work in government offices on the city, state, and federal levels. Non-profit organizations, both small firms and large agencies, employ technical writers as well.

Technical communication is performed by all business professionals. However, technical writers are professionally trained to communicate in the workplace. Technical writers are men and women who have been educated and trained past the high-school level in a professional communication program. Technical writers may have a two- or four-year degree from this type of program. Today, more universities are offering degrees in technical communication at the master's and doctoral levels as well.

In its brochure, *Careers in Technical Communication*, the Society for Technical Communication (STC) identifies nearly 160 colleges and universities offering programs and courses in technical communication. This figure does not include the fourteen programs being offered internationally in countries such as France, the Netherlands, and Canada. STC, a professional organization "dedicated to advancing the arts and sciences of technical communication," is 23,000 members strong and is one of the largest organizations of its kind in the world.

Technical writers are generally individuals who enjoy reading and writing, want to know a variety of people in different professions, and work well with others. In order to be successful in their jobs, they must be organized, conscientious, and efficient when working under deadlines.

Technical writers work on a number of projects that may take days, weeks, or months to complete. Working with a variety of colleagues—such as personnel who have technical expertise (engineers, scientists, computer programmers), designers, advertisers, printers, or other writers—is not uncommon. Technical writers must be ready to encounter a variety of subject matter because typically the topics that are being written about change with each new project.

Response Exercise 1.1

Technical writers work in many different organizations—from finance- and computer-related fields to publishing, engineering, and manufacturing. Obtain materials about an organization in your area, either directly or by accessing its Web site. The organization could be a bank, a company, or a manufacturing firm, and the materials could include an annual report, a company mission statement, or several sales brochures about the product(s) or service(s) it offers.

To analyze how technical communication is at work in this organization, read through the materials you have obtained and write down your responses to the following questions:

1. What product or service does the organization sell?
2. What do I know about its reputation? Have I ever purchased the company's product or used its service?
3. Based on what I know about the organization, what role might technical communication play there? (Is the organization a large one, necessitating a

great deal of technical communication? Does its product or service require a lot of technical communication?)

4. Based on the documents that I have accessed about the organization, does technical communication seem to be considered important by the company itself? What led me to this conclusion?

What are the products of technical communication?

Based on their work experience and where they are employed, technical writers produce a wide variety of print documents, whether they be on-line resources, such as Web sites, or oral presentations, such as those given at meetings, conferences, or with clients.

Typically, though, there are approximately seven different types of products that a technical writer may be asked to produce. While these seven types are not an exhaustive list of all products that technical writers create, they do represent the more common products, and they illustrate the range of items writers produce. All seven of these types are described in greater detail in Chapter 3. However, to introduce technical communication to you, these seven types are briefly described below.

- Marketing and informational materials
- Correspondence
- Instruction and user manuals
- Oral presentations
- Proposals
- Reports
- Web sites

Marketing and informational materials

Marketing and informational materials include a wide range of documents that inform and persuade a variety of readers. Typically, these types of materials are produced for customers or clients who are interested in purchasing your company's product or using your company's service.

The following story boxes—beginning here with the informational booklet used to describe an early American quilts exhibit—illustrate the ways in which technical communication can play a role in all sectors of professional and personal life. These story boxes describe how seven different types of technical communication—informational materials, correspondence, instruction manuals, oral presentations, proposals, reports, and Web sites—can influence a variety of career fields.

Correspondence

Letters and memos, including both print and electronic mail (e-mail) versions, are types of correspondence. Generally, letters are written to readers outside of your organization (clients, customers, vendors, colleagues), while memos are written to readers within your company (co-workers, supervisors).

Technical communication reveals the historical and artistic significance of quilts...

A seven-page informational brochure, "Early American Quilts," accompanies a local museum exhibit that features nearly two dozen quilts from the nineteenth and twentieth centuries. The booklet, a type of technical document, not only features photographs of the most important quilts on exhibit, but also describes the quilting techniques that qualify these artifacts as technically impressive creations. This booklet presents the range of quilts on exhibit, while showcasing their historical and artistic importance.

Instruction manuals

Instruction manuals tell readers how to operate and maintain products or equipment and how to perform operations and use services. In all cases, instruction manuals are written for the user, whether that reader is the user of a product, a piece of equipment or hardware, or a service.

Oral presentations

Unlike a print document, an oral presentation provides its audience with information that is spoken aloud. While presentations may include print documents in the form of hand-outs or visual aids on a projector screen or a flip-chart, they also rely heavily on the spoken word to communicate information.

Proposals

Proposals are a type of document written before work is performed that detail the work that needs to be done. Typically, proposals are written for audiences outside of your organization. For example, many engineering and construction companies send out proposals to other organizations to bid on or propose work for a job. (Note: Chapter 3 describes a type of proposal that is written by you for an audience within your organization, which is becoming more common in a variety of workplaces.)

Technical communication makes automobiles safer...

A letter being distributed by an automobile manufacturer to nearly 50,000 automobile owners describes a newly discovered safety concern regarding the passenger-side airbag. While the automobile manufacturer distributed a press release to major media about this safety concern, the manufacturer believes that the letter is a more effective communication device. The letter, a type of technical communication, not only clearly and accurately describes the safety concern in language that a typical consumer can understand, but also describes what action the car owner must take to replace the faulty air bag.

Technical communication extends the life of a lawn mower...

An instruction manual accompanying a four-cycle, self-propelled lawn mower describes how to operate and maintain the equipment. The manual not only clearly and accurately describes how to safely operate the mower, but also indicates how to maintain the equipment. Equipment maintenance extends the life of the lawn mower and allows the user to operate the mower more safely.

Proposals are evaluated according to a *technical* description of work, a description of project *personnel, budget* requirements, and *schedule* parameters. The technical component describes how the work will be performed, usually explaining the best method for getting the job done. The personnel component identifies the project supervisor and workers who will work on the project. All financial and budgetary estimates, including labor, materials, and equipment, are given in the budget component. Proposals typically contain a scheduling component that estimates the time it will take to complete the project and the length of time of each phase of work.

Reports

Like correspondence, reports communicate a wide variety of information. For the most part, reports describe various aspects of work that has been or needs to be completed. Reports are written for both external audiences (clients or other personnel outside of your organization) and internal audiences (supervisors, managers, or other colleagues within your organization).

Web sites

Web sites are electronic communication resources that can be read and used by anyone who has access to a computer and the World Wide Web. More organizations are using the Web as a way to inform potential or current customers about their products or services or to sell customers that product or service directly.

To create these on-line marketing and sales opportunities, companies create Web sites. For example, a large seed production and sales company based in the Midwest has an elaborate Web site that sells its products nationwide. The company sells a variety of seeds and seed products through its Web site. The user simply chooses the geographical region of interest—for instance, the Midwest, the Southeast, the Northwest—and a product list for that region appears.

Technical communication makes mammography safer...

An oral presentation, being given to twenty radiology department staff members, describes the new mammography patient record-keeping procedure. By law, this procedure will become the new standard for radiology department staff to follow. The presentation not only clearly and accurately describes the new procedure to the staff, but also illustrates what the consequences are if the procedure is not followed.

Response Exercise 1.2

Technical writers create many kinds of products, from documents, such as reports and proposals, to oral presentations and Web sites. Obtain materials about an organization in your area from which you do business, either directly or by accessing its Web site. The organization could be any type of non-profit agency, company, or manufacturing firm. The materials included could be an annual report, a company mission statement, or several sales brochures about the product(s) or service(s) offered by the organization.

To analyze the types of technical communication used at this organization, read through the materials and write down your responses to the following questions. (Note: If you have accessed the company's Web site, click on various pages of the site to identify individual "documents," such as a mission statement, company history, or product/service information. Then, respond to the questions based on three or four of these "documents.")

1. What kind of documents did I obtain? In two to three sentences, describe each document. For example, what kind of information does each document contain?
2. Who is the intended audience for each document: a customer, a shareholder, a potential client? In each document, what assumptions—about gender, age, income level, interests, race, ethnicity—are being made about that audience?
3. Taken together, what do these documents say about the company? For example, what kind of reputation do these documents create for the company: respectable, trustworthy, long-standing, dependable?
4. From your prior knowledge of or experiences with the company, is this reputation valid? If not, how do your experiences differ from the reputation the documents suggest? If so, has your exposure to the company's technical communication—the documents you are reading now or other kinds—influenced your beliefs about the company?

Technical communication creates work (and makes money) for companies…

A multi-million dollar proposal, written to bid on the construction of eleven fast-food restaurants, describes the technical, financial, scheduling, and personnel components of the proposed work. The proposal is one of several that are received at the fast-food restaurant chain's home office. The proposal not only proposes how the work will be completed, but also attempts to sell the long-standing reputation of the construction company.

Technical communication describes important trip details…

A trip report, describing a business trip taken to Canada, informs colleagues and supervisors about the purpose of the trip, the work performed, and the accomplishments of the trip. This trip report describes important progress made with the supervisors in the company's Toronto-based headquarters. The trip report, a type of technical communication, not only describes the tasks accomplished during a two-hour meeting with these supervisors, but also illustrates the supervisors' reactions and ideas in detail.

What are the characteristics of technical communication?

Technical communication is comprised of five characteristics that you must understand if you are to produce effective technical communication. Successful technical writers understand that, with every document they create, they must concern themselves with these five characteristics because they know that these characteristics impact the quality of technical communication.

- Communication context (situation, purpose, audience)
- Content
- Organization
- Prose style
- Design

Technical communication promotes tourism...

A Web site describing six lakeside vacation cabins informs potential vacationers about the location, condition, amenities, availability, and price of the cabins. This Web site shows the beauty of the lakeside location by including photographs of the cabins and the surrounding area. The Web site not only helps to sell the lakeside cabins, but also helps to promote attractions in the nearby community by providing links to these sites as well.

Since documents are always linked to actual events, tasks, circumstances, and people, the **communication context** of a document refers to the *situation* that necessitates that document, the *purpose* for the document, and the *audience* for or readers of that document. The **content** of a document is the subject matter or topic of that document. A document's **organization** refers to the way in which its information is ordered and arranged. The **prose style**—or writing style—should be clear, concise, and courteous. The **design** of the document is the visual layout of the document on the page.

These qualities are referred to throughout the rest of the book and discussed in detail in Chapter 3. To see how all of these characteristics play out in a document, read the document scenario below.

"Special Spring Flowers" document scenario

You work as a technical writer for a fresh flower distributor that delivers fresh flowers to its chain of small retail flower shops across the country. Your supervisor introduces the *situation* of your next project by asking you to revise "Special Spring Flowers," the full-color, 8 1/2" × 11" booklet that illustrates and describes the flower arrangements showcased and "on special" during the spring season (roughly mid-April through mid-June). Your supervisor tells you that since the booklet must be ready for the printer by the first of the month, you need to begin work immediately.

This booklet is distributed to all retail shops as a way for shop employees to show the booklet's primary *audience*, shop customers, the arrangements that are available for the season. Several shop employees, the booklet's secondary *audience*, have complained about the booklet's poor organization and inaccessible design.

To begin planning how you will revise the current booklet, you read through it carefully. You note that the primary *purpose* of the booklet is to inform customers of the variety of flowers they may purchase. Each arrangement has a title, a paragraph-length description, a list of flowers used, and a full-color photograph. These details—plus a price chart—make up the *content* of the booklet.

To revise the booklet, you decide to take the suggestions of shop employees and focus on improving the organization and design of the document. To create a more customer-friendly *organization*, you order each flower arrangement in the booklet by price range. Then, you create a chart at the beginning of the booklet that includes small photographs of each type of flower available, along with *concise, clearly written* flower descriptions. The customers can look at the chart, identify one or several flowers that they would like in an arrangement, and locate which arrangements being offered feature those flowers.

You make the *design* of the booklet more appealing by striving for consistency among the font styles and sizes. Also, you use the yellow, blue, and green colors found in your company's logo as color themes to tie together the "look" of the booklet.

For each document you create as a technical writer, you must consider the five characteristics of technical communication illustrated above. Creating documents, or other resources such as Web sites or oral presentations, takes attention to detail, knowledge of communication, and an ability to manage these characteristics.

Response Exercise 1.3

Understanding the common characteristics of effective technical communication is important. From an organization in your area, either obtain a document (an annual report, product or service brochure, correspondence), or access its Web site. The organization could be any type of company, non-profit agency, or manufacturing firm in your area.

Technical communication exhibits five characteristics: communication context (situation, purpose, audience), content, organization, prose style, and design. To analyze these characteristics in the document you have chosen, read through the document and write down your responses to the following questions. (Note: If you have accessed the company's Web site, click on various pages of the site to identify individual "documents," such as a mission statement, company history, or product/service information. Then, respond to the questions about this "document.")

1. Who is the intended *audience* of this document: a customer, a shareholder, a potential client? What elements in the document suggest that it is geared toward this audience?
2. What is the *purpose* of the document? Is the document specifically to inform, persuade, report, recommend? Can you speculate anything about the *situation* surrounding the creation of this document?
3. Is the *content* suited to the audience? Would the audience find it interesting?
4. Does the *organization* of the document allow the audience to understand the content more effectively and efficiently?
5. Does the *prose style* suit the message of the document? Does this style help or hinder your understanding of the message?
6. Does the document's *design* facilitate audience use? Is the content made more or less accessible by the design?

Who benefits from effective technical communication?

All organizations, regardless of their size and profit margins, rely on and benefit from effective and efficient technical communication. In very general terms, an organization benefits by increasing its customer base, building and maintaining a respected business reputation, and facilitating more efficient, less costly work operations.

Organizations increase their customer bases by sending out their product and service information to potential customers in effective ways—through a well-designed Web site, usable product and service brochures, and accurate and user-friendly product instruction manuals, among other things.

Professionals working in organizations that regard effective technical communication as a high priority also benefit. Professionals such as engineers or other personnel with technical expertise rely on well-crafted technical communication, created by themselves and others, to describe and communicate the work that they propose, are performing, and have completed.

Supervisors or managers—at any level and within any department in an organization—benefit from ready and efficient communication with colleagues who may be located within or outside of their departments, with clients at other organizations, and with potential and current customers. Technical communication—from correspondence to annual reports to marketing information—helps these professionals communicate.

Consumers benefit from technical communication as well. Finding and using a well-organized and engaging Web site helps consumers gain a wealth of product or service information. Consumers who purchase products with accessible and accurate instruction manuals benefit from understanding the products' operation and knowing the correct safety precautions.

Chapter 1 Task Exercise

This Task Exercise asks you to conduct the first of a four-part interview with a technical writer. Identifying and describing the importance of technical communication at this writer's workplace is the topic of this interview. (Task Exercises in Chapters 2–4 describe the remaining three parts of the interview.) From these interview responses, you will create a Workplace Communication chart that illustrates the different types of technical communication produced at the technical writer's organization.

Select your subject: To prepare for the interview, select your interview subject. This subject should be a full-time technical writer at the organization of your choice. He or she could be a friend or someone you have worked with, or even someone you have never met before. (Note: Professionals are generally willing to help those interested in learning more about their careers, so do not be afraid to ask someone you do not know.)

Prepare for the interview: Once you have identified a possible interview subject, contact him or her on the telephone or write a letter asking for an interview. It's important to be prepared to clearly and concisely describe the purpose of your call. Using *the interview purpose statement* will help you to describe the purpose and uses of the interview to the subject.

To prepare the interview purpose statement, write down (a) the purpose of the interview, (b) the use you will make of the responses you receive, and (c) the approximate amount of time you will need for the interview. (It is probably wise to conduct the interview in either four or two parts, each part lasting roughly half an hour.) When you request and schedule your interview, make sure you communicate these three points to your interview subject. Once you have scheduled the interview, prepare and then review your interview script (the series of questions you will ask the interview subject).

Interview questions: Ask the writer the following questions about the nature of technical communication at his or her workplace. The responses you receive will help you to create your Workplace Communication chart.

1. What types of technical communication do you produce (marketing and informational materials, correspondence, instruction manuals, oral presentations, proposals, reports, or Web sites)? What types do you produce most often?
2. For each type, identify the five characteristics of effective technical communication (communication context, content, organization, prose style, design).

3. For each type of communication, name the people who benefit from this kind of technical communication (customers, clients, colleagues, supervisors). Why and how do they benefit?

During the interview: Be sure to arrive on time at the selected location and come prepared. When you arrive for the interview, make sure to introduce yourself and summarize the interview purpose statement. Also remember to bring along a *watch* to make sure you keep the interview within the predetermined time, a *notepad* and at least two *pens* to take notes, and a *tape recorder* to record the interview (ASK PERMISSION FIRST). Conclude the interview by thanking the individual, and be sure to request a follow-up interview, if necessary.

Interview follow-up: After you have conducted the interview, send the individual a thank-you note. Also, if a follow-up interview is necessary, either in person or via telephone, be sure to conduct the interview in the same courteous manner that you did the initial interview.

Workplace Communication chart: From the interview, you have received a wealth of information about the uses, nature, and benefits of technical communication in one workplace. From this information, create a Workplace Communication chart that identifies, organizes, and describes the following information:

- Identifies the name, position, and years of experience of the technical writer you interviewed.
- Identifies and describes the workplace you visited (the products and services offered there).
- Identifies the types of technical communication used at the workplace.
- Describes, through a discussion of the five characteristics of technical communication, each type of technical communication.
- Provides an annotated list of who benefits from this technical communication and why.

You may organize and design your chart in any way—as long as it communicates these points in a usable and appealing way. For example, you may use poster board to create a large and colorful chart, or you may use a software application to create a chart that can be easily saved, edited, and revised.

Chapter 2

What Do Technical Writers Do?

To understand the varied and interesting work that technical writers perform, you must identify where and with whom they work, what kinds of products they create, and what education and training are necessary to begin a career in the field of technical communication.

Because the communication skills that technical writers use every day are more highly in demand now than ever before, technical writers may choose to work nearly anywhere in the country or world. The variety of arenas in which technical writers can work is virtually unlimited, so whether you are fascinated by computers and the World Wide Web, by international business communication, or by the manufacture of cutting-edge technology, you're bound to find your niche.

According to the *1998 Technical Writer Salary Survey*, published by the Society for Technical Communication (STC) (one of the largest professional organizations in the field), salaries for technical writers and editors rose 3.2 percent from 1997 to 1998. The *Survey* finds that the average salary for entry-level technical writers and editors is now $36,100. These figures prove that the demand for technical writing skills is on the rise.

To help you to recognize whether or not a career in technical communication is right for you, this chapter asks the question, "What do technical writers do?" This chapter answers that question by discussing where and with whom writers typically work, what they produce, and what training and education is necessary to become successful in this field. Later in this chapter, you'll read profiles of two experienced technical writers. You'll learn about their training and current work, and you'll see firsthand what it takes to be a technical writer as well as the variety of available career options and experiences that the field of technical communication offers.

After you have read this chapter, you will know the answers to these basic and important questions about the elements of a career in technical communication:

- *Where* do technical writers work?
- *With whom* do technical writers work?
- What kinds of *projects* do technical writers perform?
- What *education* and *training* are necessary?

While this chapter does not present all of the tasks that a technical writer might perform on the job, it does give you—someone interested in technical communication as a possible career—a valuable introduction to this growing and diverse field.

Where do technical writers work?

The brochure, *Careers in Technical Communication*, published by STC, notes that the 23,000 members of its organization are "employed in every industry, from automobiles to computers to finance." The diversity of industry types from which technical writers may choose is one of the most attractive aspects of the field of technical communication. Depending upon your interest, you may choose a position in a manufacturing, computer, business, finance, or non-profit-related field.

Technical writers are employed by large companies in cities in the United States and abroad in Europe and Asia. They are also employed at smaller companies in both urban and rural areas of the country.

Because technical writers are employed at a variety of types of companies, the work that they perform is varied as well. Table 2.1 shows the different categories of industry in which technical writers are employed, and it includes a brief list of writers' main tasks and responsibilities.

While these descriptions of work and responsibilities are brief, they give a good indication of the main tasks for which technical writers are responsible. Also, each workplace environment creates subtly different types of tasks and responsibilities. The stories of two technical writers, Stan and Joan, who we introduce in the next section, help to illustrate the different types of workplaces at which technical writers are employed.

Two technical writers—Two different companies

Stan Cho: As a technical writer, Stan Cho works in the computer industry as a project lead/senior staff analyst at *TechnoInc Corporation*, a software and professional services consulting company. Holding a master's degree in professional communication, Stan has been working with TechnoInc for nearly one year.

Table 2.1 Categories of workplace environments and work and responsibilities

Workplace Environment	*Typical Work and Responsibilities*
Manufacturing • Create documentation to support plane, auto, or other equipment manufacture. • Work with engineers, designers, information systems personnel, and other writers.	• Create systems documentation or user documents, such as instruction manuals. • Maintain document record of equipment design and manufacture.
Computers • Help build e-commerce businesses, create computer software and hardware, and design Web sites. • Work with computer engineers, programmers, and software development teams.	• Create systems documentation or on-line support documentation, such as tutorials or help lines. Design and build Web sites. • Help e-commerce businesses become more competitive.
Business and Finance • Build and maintain communication network among colleagues, clients, and customers. • Work with information systems personnel, business managers, supervisors, and legal personnel.	• Create a variety of on-line and print documents, such as newsletters, reports, information, sales, and other promotional materials. • Create and maintain organization's communication network.
Non-profit • Build and maintain communication network among colleagues; clients; and those who donate time, money, or services to your organization. • Work with the public, supervisors, clients, and volunteers.	• Create a variety of on-line and print documents, such as newsletters, reports, information and marketing materials. • Create and maintain organization's communication network.

TechnoInc Corporation is a large company that has been in business since the early 1970s. TechnoInc's home office, located in Michigan, has an impressive client base that continues to grow—TechnoInc's software renewal rate is nearly 100 percent. TechnoInc helps organizations optimize their software needs by helping to design, build, test, and maintain the types of applications that businesses use to run efficiently and effectively.

The following story boxes—beginning here with the description of the technical writer employed in the manufacturing industry—illustrate the kinds of work technical writers perform in different businesses and industries. These story boxes describe five industries in which technical writers are employed—manufacturing, computers, business, finance, and non-profit—and the kind of work writers do in each.

A technical writer working in the manufacturing industry...

As a technical writer, you work for a multi-national corporation that designs and builds earth-moving equipment (bulldozers, front-end loaders). You collaborate with other writers and information systems personnel to create reports—containing elaborate hardware descriptions of the equipment—for clients interested in purchasing your corporation's product. Through the reports you create, clients understand the capabilities, functions, and uses of the equipment of interest to them. Completing reports like these gives you, as a technical writer, the satisfaction of being a part of designing, building, and selling the equipment that helps to build our world.

Joan Reese: As a technical writer, Joan Reese works in the manufacturing sector as a functional analyst at *Flite Manufacturing*, one of the largest aerospace companies in the world. Joan moved to the Northwest to work at Flite Manufacturing's home office in Washington state after having completed a master's degree in Business and Technical Communication. After two years at Flite Manufacturing, Joan continues to work at new and challenging positions within the organization.

Flite Manufacturing manufactures commercial and military aircraft, including helicopters, electronic and defense systems, missiles, rocket engines, launch vehicles, and advanced information and communication systems. While based in the Northwest, Flite Manufacturing has plants throughout the country, with clients that include the United States military. This multinational company has customers in over 140 countries and employs over 200,000 people.

As you can see, technical writers, such as Stan and Joan, work in very different workplace environments. In the next section, you will see how the environments within which Stan and Joan work—a computer software corporation and a large manufacturing company—also affect how frequently, with whom, and how they collaborate.

Response Exercise 2.1

The diversity of workplace environments technical writers have to choose from is virtually unlimited. Go to your public library and find out more information about where technical writers work. Find recent articles in newspapers, such as the *New York Times*, or in magazines, such as *Fortune, Forbes, U.S. News and World Report,* and *Business Week,* that describe careers in technical communication.

To discover where technical writers work, read through the materials you have obtained and write down your responses to the following questions.

1. What are the most popular types of organizations that employ technical writers? Manufacturing? Computers? Publishing?
2. Choose one organization from this list and describe it. What types of products does it create, or what services does it provide? How do technical writers help this organization do business?
3. What role do technical writers play in international business? Do technical writers have career opportunities in other areas of the world, such as Europe, Asia, or South America?
4. What criteria do you have for a corporation where you would like to work?

With whom do technical writers work?

While people who are not familiar with the writing process may believe that writing is best performed individually, most of the writing that occurs in the workplace is performed collaboratively—that is, in a team.

Figure 2.2 shows the stages of the writing process. Writers can expect to collaborate with other writers, information systems personnel, and personnel with technical expertise (such as engineers, scientists, and computer programmers) throughout this four-stage process.

Figure 2.2 The four stages of the writing process

Planning ↔	Drafting ↔	Revising ↔	Editing
• Identify communication context • Brainstorm/locate more information • Create visual aids • Make document design decisions	• May use outline or notes to help generate text. • Continually makes communication context, content, organization, prose style, and design decisions. • May collaborate with others to help generate text.	• Revise for accurate and complete content • Revise for audience • Revise for clear and logical organization • Revise for proper mechanics, prose style	• Edit for accurate and complete content • Edit for logical organization • Edit for accurate and consistent document design • Edit for mechanical, prose style, and voice errors

A technical writer working in the computer industry...

As a technical writer, you work for a small, Web-based computer consulting agency that helps other companies build and design their Web sites. This week, you work with a client and with a colleague in the sales field to build that client's Web site. The client—an art gallery owner who displays and sells local artists' work at his main street gallery—wants a Web site that displays artwork currently on exhibit, while enabling interested buyers the opportunity to purchase art through the Web site. As a writer, completing this project is satisfying because the Web site you have helped to create is not only usable and aesthetically appealing, but also supports talented, local artists by helping them to gain recognition and sell to a much broader audience.

As a technical writer, the kinds of projects to which you contribute will depend largely upon the type of work your company does. However, regardless of the industry you're in, expect to learn a great deal about other fields—such as engineering, science, computers, and design—and to work with the people who work in those fields.

To illustrate the collaborative nature of two organizations, we describe the work environments of both Stan, a technical writer for TechnoInc, and Joan, a technical writer for Flite Manufacturing. The next section illustrates the kinds of collaboration both perform and how these two writers perceive their workplaces.

Two technical writers—Two different work environments

As a senior staff analyst at TechnoInc, Stan writes a variety of documents, including consulting proposals and marketing materials. Mostly, though, Stan writes and designs electronic resources, such as Web sites, Web-based applications, and Web-enabled databases.

Stan notes that "As a consultant, I have the opportunity to work on a variety of projects with many different people in different companies." Depending on the nature of the project, Stan may work with Web developers, project managers, software engineers, writers, designers, programmers, data modelers, database administrators, and clients who may include marketing, sales, and IT (information technology) professionals. Stan also indicates that the collaborative experiences that he has with his clients and colleagues are some of the most positive aspects of his career as a technical writer.

As a functional analyst for a large manufacturing corporation, Joan collaborates most closely with programmers and information systems personnel. Joan notes that at Flite Manufacturing, facilitating collaboration is an important aspect of creating a successful work environment. Therefore, one of the primary objectives of developing a large, corporation-wide intranet was to enable employees to work more efficiently and effectively together.

On a recent project, Joan encouraged the multiple computing support organizations within the company to work together to fill factory computing needs.

Joan states that while the importance of collaboration is a relatively recent corporate mission, it is an important one for Flite Manufacturing. Flite Manufacturing hopes to enable collaboration on all levels—from integrating product teams to building supplier and customer relationships.

Since Joan and Stan work in different environments, they collaborate with different types of colleagues or clients. However, it is important to realize that for both writers, collaboration is an integral part of their work. Successful companies, such as TechnoInc and Flite Manufacturing, are becoming increasingly aware of the importance of facilitating collaboration on all levels of the organization—among colleagues, supervisors, clients, and customers.

What kinds of projects do technical writers perform?

The kinds of documents and other resources that you create as a technical writer depend largely on the products and services that your organization offers. For example, a technical writer working for a large automobile manufacturing corporation may create different documents than a writer who works for an airplane manufacturer. Technical writers in a manufacturing sector can expect to create different kinds of products than those writers working in the finance or non-profit sectors.

However, even the products of technical writers working in the same organization may be very different. For example, writers working within an automobile manufacturing corporation work on different projects according to their locations within specific departments or divisions: a writer in the safety division produces different types of documents than the writer who works in the research and development division.

A technical writer working in the business sector…

As a technical writer, you work for a large architectural firm that designs both commercial and residential spaces. On a current project, you work with a marketing colleague and one of your firm's partners to create a new informational booklet for your firm. The booklet is designed to inform potential clients about your firm's available services, the qualifications of its personnel, and its past design projects. As a method of building your firm's client base, this booklet must illustrate that your firm places the needs of its clients first. The most satisfying aspect of completing this informational booklet is that as a technical writer, you help demystify the process of hiring an architect for the client. The booklet may even convince hesitant homebuilders that their projects are important enough to warrant your firm's professional services.

It is worth noting, however, that the products technical writers create—regardless of their corporation or department—are similar in several important ways. Each document and resource that writers produce is first planned by attending to its communication context (situation, purpose, audience), content, organization, prose style, and design. (Note: For discussions about these five characteristics of technical communication, see Chapters 1 and 3.)

Technical writers are involved in a variety of diverse projects. As a technical writer for TechnoInc, Stan works collaboratively with his colleagues and clients. As a technical writer for Flite Manufacturing, Joan collaborates with her colleagues as well. The next section discusses the kinds of projects these two writers produce and the different—yet strikingly similar—ways that each writer works on a project.

Two technical writers—Many types of projects

Technical writers produce many different types of documents, both print and electronic versions, and other communications, such as oral presentations. For example, during his employment with TechnoInc, Stan has helped to create a variety of electronic resources, such as Web sites, intranets, extranets, and print documents, such as statements of works, consulting proposals, marketing materials, project documentation, and a business plan.

Flite Manufacturing hired Joan as a contract technical writer to document processes in flow chart diagrams for a company-wide reengineering effort. In this project, Joan's role was to document *current* processes versus *desired* processes within a given area of Flite Manufacturing's manufacturing/business cycle.

Later in her career, working on another project (and with a new title), Joan collaboratively planned, designed, and prepared presentations to illustrate how *factory* computing needs are different than traditional *desktop* computing needs. These presentations helped to encourage the many computing support organizations throughout the company to work together to support factory computing needs.

Currently, as a functional analyst, Joan provides guidance and training in the use of applications, data files, and databases that are used to retrieve and manipulate data. On a current project, Joan writes functional specifications for extensions to a software program. Initially developed in-house, the program is now used worldwide by Flite Manufacturing. Flite Manufacturing uses the program at customer sites during meetings with airline customers when airplanes are being configured for order.

While both the work Joan and Stan perform and the products they produce are different, both writers move through the writing process in relatively similar ways. Read through the following section to discover the similarities between the way Stan creates a statement of works document and the methods Joan uses to construct an oral presentation.

Two technical writers—How do they write?

Stan began his statement of work document by selecting a template to help him plan the document. Stan continued the writing process by working collaboratively with a team of colleagues. While the document was drafted and revised collaboratively, Stan was the primary author, and he made the final decisions about both making substantive changes in the document and implementing suggested feedback from his team members.

A technical writer working in the field of finance...

As a technical writer, you work at the home office of a bank that has branches throughout the state. For this project, you create a set of documents to be distributed to all branch banks in an effort to promote its 5.95 percent residential mortgage loan campaign. The documents, geared toward current and potential customers, include a poster to be placed in each branch's lobby, an informational brochure to be inserted with every current customer's monthly bank statement, and a one-page flyer to be included as an insert with the local newspaper. Working closely with the printer, you create a set of interesting, informative, and aesthetically appealing documents. The most satisfying aspect of completing this project is knowing that the documents not only convey all the important information about the loan campaign, but that they also communicate this message in a dynamic, engaging way.

In the latter stages of the writing process, Stan's team members read through the statement of works and edited it for accuracy, appropriateness, and appeal. The changes recommended by each team member were indicated on a printed copy of the statement of works. Stan read through the comments and suggestions and incorporated the necessary changes.

After four drafts of the document, Stan and his team had a statement of works document that fulfilled the necessary qualities of technical communication: communication context, content, and organization, along with the specifics concerning prose style and design.

Joan's product, a presentation given to Flite Manufacturing factory teams at multiple locations, provided follow-up to a week-long working session. The audience was comprised of computing service providers and managers. Like Stan, Joan drafted and revised the presentation collaboratively, and Joan was the presentation's primary author. Not only did two other team members provide Joan with input and editing comments, but she received feedback from one member of the project's management team.

Joan began planning the presentation by formulating the message she wanted to convey. Specifically, Joan wanted to remind the factory team audience of the project's purpose, the great results of the week-long session, and what to expect next.

After sketching an outline on paper, Joan began formulating headings in the presentation template she had used successfully in the past for other presentations. The bulk of Joan's work included drafting the presentation's bulleted comments and points. After she fine-tuned the wording, Joan asked her two teammates to review the draft of the presentation. Using their suggestions and comments, Joan incorporated nearly all of their feedback into the presentation.

Since Joan gave this presentation many times, she continued to revise the wording and pace of her delivery while elaborating more on certain points along the way, as necessary. Joan also revised the presentation "on the fly"—as she delivered it to her audience—by picking up the pace if her audience seemed groggy, slowing down the pace if they looked confused, or elaborating on points that seemed of most interest to them.

Response Exercise 2.2

Technical writers work with many different types of people to produce a variety of documents. To understand what your co-workers can contribute to a document—either during the early planning stages or during the latter revising stages—you must know what kinds of work your colleagues perform. By knowing your colleagues' tasks and responsibilities, you can help to gauge when and how their input will be most valuable.

Some of the types of personnel you will work with include designers, printers, engineers, information systems managers, computer programmers, and software engineers. To understand the work these professions entail, locate a jobs guide or a career atlas in your local library.

To learn more about the work that your colleagues perform, make a list of those personnel with whom you may work (include any or all members of the list above), and use the materials you have obtained at the library to locate information about each colleague. Then, write down your responses to the following questions:

1. What work does each colleague perform? What are the main job duties of each colleague? What are the main job responsibilities?
2. What education does each colleague require? What training is necessary?
3. What tools or equipment must each colleague know how to use? Computer? Auto-cad? Design equipment?
4. How important is communication with each colleague? What role does communication play with each colleague?
5. In what ways could each colleague help you to write, design, or revise a document?

What education and training are necessary?

Technical writers are men and women who have been trained and educated past high school in a professional communication program. While both Joan and Stan have their master's degrees in professional communication, many technical writers are employed with a two- or four-year degree from this type of program.

In its *1998 Technical Writer Salary Survey*, STC reported that solely in terms of education level, the median salary for a technical writer with a bachelor's degree is $47,300, while the median salary for a writer with a master's degree is $49,460. To find out more about the education and training requirements for technical writers, read Chapter 4, which discusses how to choose the right type of professional communication program for you and describes several popular programs.

Next, we briefly summarize Stan and Joan's education and training and discuss what else was needed to prepare them for their respective careers in the computer and manufacturing industries. Also, Stan and Joan offer insights about their work environments and advise those interested in joining the technical communication field.

A technical writer working for a non-profit organization...

As a technical writer, you work at a local non-profit organization that provides information to those interested in understanding more about Alzheimer's disease. Your next project is to produce an oral presentation script and a corresponding set of transparencies to be used by your organization's volunteers during information meetings and lectures. The script and transparences are used by volunteers to provide information to their audiences about recognizing the differences between Alzheimer's disease and mental illness. Working closely with the organization's supervisor and a registered nurse, you create a set of complete and informative documents that any volunteer can use, whether new to or familiar with the program. Completing this project is satisfying because you know that these easy-to-use documents are helping concerned citizens correctly recognize the symptoms of Alzheimer's disease.

Two technical writers—How did they train and prepare?

Both Stan and Joan received their master's degrees in professional communication, and both had internship experiences before their current employment. In terms of preparation, Stan and Joan offer the following advice to anyone thinking about a career in the field of technical communication.

Stan stresses the importance of developing a wide range of skills: "Foster analytical thinking, while developing leadership qualities. Be sure to stay abreast of industry, market, technological, and hiring trends. One relevant, practical suggestion that I would give to others interested in the field of technical communication is to understand how the Internet is changing the world of business and how it is changing the professional lives of technical writers."

Joan notes that choosing to work for a large company has both positive and negative aspects. For example, the benefits, such as off-hours training, paid continuing education, and a choice of location, are more available in a large company. Joan notes that her flex schedule option is a benefit as well. Joan works eighty hours in nine days and takes off the tenth day (every other Friday). Despite these benefits, though, Joan dislikes her "cube farm environment"—cubicles with very little privacy—which is not as aesthetically pleasing as the office with a window that she may enjoy at a smaller company.

Joan also states that it was important for her to work for a company that produces something substantive, and although she is "not riveting together a 747," her work brings about something at the end of the day.

Joan notes that in the field of technical communication, it is quite important to "be flexible." Also, she notes that as a technical writer, "you may be in many different companies and industries over the course of your career—keep an open mind about each and the options they hold. You will be expected to learn extensive amounts about the industry you enter. Be willing to dive in, take classes, make friends from different backgrounds, and 'assimilate' so that you are effective in your current area."

Response Exercise 2.3

Technical writers work in many different organizations—from finance- and computer-related fields to publishing, engineering, and manufacturing. Do an online search for technical communication jobs, or obtain a listing of several technical communication jobs from local newspapers.

To analyze the kinds of skills you will need, read through the materials you have obtained and write down your responses to the following questions:

1. What types of organizations are seeking help? In three to four sentences, describe the products or services these organizations offer.
2. What computer skills—word processing, Web design, programming—are listed most often? Why?
3. What specific communication skills—writing, editing, design—are listed most often? Why are these skills important?
4. What other skills are listed frequently: collaborating, working on deadlines, frequent traveling?

Besides thinking about the type of professional communication program that is right for you, it is important to know what you want from a career and if a career in technical communication is suitable for you. To help you make a decision, prepare the following Task Exercise. It asks you to interview a working technical writer about the aspects of his or her job. Besides simply interviewing the writer for this exercise, you may ask to tour the writer's corporation, or you may request to "shadow" the writer for one work day. Performing a "job shadow," in which you follow the writer during a normal work day, gives you the opportunity to ask the writer questions, see the writer in action, and listen while the writer works with colleagues, clients, and customers. This type of job shadowing is a valuable way to begin to understand whether or not technical communication is the career field for you.

Chapter 2 Task Exercise

This Task Exercise asks you to conduct the second of a four-part interview with a technical writer. Identifying and describing the work he or she performs is the topic of this interview. (Task Exercises in Chapters 1, 3, and 4 describe the remaining three interview parts.)

Interview a technical writer about his or her education and training experiences, past and current work experiences, experiences working with collaborators, and advice for those interested in technical communication as a career. Using the interviewee responses, write a two- to three-page letter to the career counselor at your local high school describing your findings.

Begin this exercise by devising a script of interview questions to ask your subject. These questions must attempt to find out as much as possible about the work that this technical writer performs. Listed below are several questions you may choose to include.

1. What company do you work for? What kind of work does your company do?
2. What is your job title? How long have you worked for this company?
3. What education/training do you have? (College degree, on-the-job training, both?)
4. What kinds of documents and other resources—for example, oral presentations, on-line materials—do you write?
5. With whom do you collaborate: other writers, engineers, designers, marketing?
6. Identify one document that you have recently written and describe the writing process that you went through to produce the document.
7. What do you like most about your job?
8. What do you like least about your job?
9. What suggestions would you give those who are interested in pursuing a career in technical writing?

For information about how to (a) select your subject, (b) prepare for the interview (c) conduct the interview, or (d) properly follow-up the interview, see the Chapter 1 Task Exercise (page 13).

Chapter 3

What Do Technical Writers Produce?

In order to understand the work involved in a career in the field of technical communication, you must know that technical writers produce a variety of materials, such as documents, oral presentations, and Web sites.

As a technical writer, one of your main responsibilities is to help professionals in a variety of occupations—from business and manufacturing to computer-related and non-profit fields—communicate effectively and efficiently with one another, their clients, and their customers. Not only do technical writers help professionals communicate, but in doing so they enable those professionals and their organizations to become more successful.

In Chapter 2, we explored the variety of organizations in which technical writers work. In this chapter, we continue this discussion by identifying and describing the variety of documents and other types of resources, such as oral presentations and Web sites, that technical writers produce in those workplaces. While the information we present in this chapter is not exhaustive, it does provide you with a good indication of the variety of work produced by technical writers.

One of the most interesting aspects of a career in technical communication is that the work you perform each day allows you to generate a wide variety of documents for a range of audiences. In the same week, you may write several dozen e-mail messages to colleagues and supervisors regarding the progress of one or more of your projects. Then, you may work with a colleague to complete a brochure describing the correct procedure for buckling up children in car seats, and assist another colleague by editing her team's lengthy recommendation report.

This chapter will help clarify the types of products technical writers produce and the qualities that make these products effective. By the end of this chapter, you will know the answers to these two questions:

- What types of *products*—documents or other resources—do technical writers create?
- What *qualities* characterize a successful document or resource?

Producing effective documents and resources is both personally and professionally rewarding. Not only do you see your written products having a positive effect on your organization's business, but you also see how the public reacts to and uses the documents you create.

What types of products do technical writers create?

Technical writers produce a wide variety of materials—such as documents, Web sites, and oral presentations—every day. Depending on the type of organization you work for and the products or services that your company offers, the documents you create range in length, complexity, and topic.

Sales materials

Sales materials are those documents that help to sell your company's products or services. The types of sales materials that you create are as diverse as the products your company sells. These sales materials may include correspondence, one- to two-page package inserts, flyers or leaflets, posters, advertisements, booklets, and even the packaging that advertises and protects the product.

Sales materials not only help persuade potential customers to buy—they help create and maintain your company's reputation and corporate image. The materials accompanying a particular product or service go a long way toward driving their sales.

Through a series of story boxes, beginning with "Susan's story...," we describe one technical writer's recent project, identify the writer's collaborators, and provide excerpts from four different products that the writer helped to create. These story boxes illustrate the different types of documents that result from one writing project. From this project description, you can see not only the variety of products that writers create, but also the range of information those products give their audience.

Instructions

Instructions describe, through text and visual aids, how to perform a specific operation. As a technical writer, you may write instructions for operations as diverse as assembling a bookshelf, operating a ship's navigational controls, or maintaining a sophisticated piece of factory equipment.

Regardless of what you're explaining, however, you must be particularly careful to create instructions that are complete, accurate, and user-friendly. Of all the documents that you create, instructions probably benefit the most from usability testing, where a document is tested thoroughly before being read and implemented by the actual user. (Note: Read Chapter 13 for discussions regarding the most common types of usability testing.)

Susan's story...

As a technical writer working on contract, Susan MacElroy began a months-long writing project for the Crystal River Community Hospital, a mid-sized health care facility serving a largely rural population in the Midwest. Susan will collaborate with a five-member team of both medical and administrative staff on the production of informational materials geared to launch Crystal River's new Women's Wellness Program (WWP).

The topic for the WWP's inaugural March program is *mammography services and breast cancer prevention*. Susan's team is responsible for planning and presenting a one-hour lecture and question/answer session about breast cancer awareness and creating informational documents to be distributed to WWP participants during this program.

Correspondence

Like all professionals, technical writers produce a great deal of correspondence. Letters and memos (both print and electronic types) are mainstays of communication in business and industry.

Letters enable writers to contact and keep in touch with clients, customers, and other colleagues outside of their own organizations. Memos allow you, as a technical writer, to maintain contact with colleagues within your organization. Correspondence—particularly if sent and received electronically—allows you to communicate efficiently and cost-effectively, and enables you to create a correspondence archive.

Susan's WWP team consists of management and medical personnel...

Susan's WWP team is comprised of the hospital's public relations manager, Joseph Jackson, who has management and business experience, but no medical training, and four radiology department personnel with little management or business training: Jan and Norman, two registered nurses; Dolores, a medical imaging technician; and Gary, a medical physicist.

The team's March WWP is only two months away, and project and task deadlines are looming. To allow for sufficient printing time, Susan's team decides first to create the informational brochure to be distributed to WWP participants. The brochure will be passed out to those attending the team's March WWP lecture tentatively titled, "Mammography Services and Breast Cancer Prevention."

An archive, containing letters and memos regarding all aspects of your work, is a valuable resource to maintain.

Newsletters

Newsletters describe company events, functions, and individual personnel and team achievements to readers such as company employees, customers, or clients. Newsletters may be produced on a weekly, monthly, or quarterly basis, and the scope or focus of the newsletter is typically decided by the company.

Depending on their audience, newsletters may be printed on high-quality paper or on relatively low-cost stock. Also, newsletters are circulated in different ways, depending on the type of audience they reach. Employees receive newsletters with their paychecks, clients or vendors receive them directly at their companies, and customers receive them through the mail.

Proposals

Proposals help sell the product or service that either you, as a technical writer, or your organization is in a position to offer. In business and industry, a proposal is used differently in different situations. For example, you may collaborate with a team of a dozen other colleagues—such as personnel with technical, legal, or financial expertise—to write a proposal in response to a document called a Request for Proposals (RFP). An RFP sent out by Company A allows Firm 1, Firm 2, and Firm 3 to compete against each other's proposals for a chance to provide the product or service being requested by Company A. By writing a successful proposal, your company will "win the bid" to produce the product or service being requested by Company A.

However, responding to an RFP is not the only way a technical writer may write a proposal. For example, as a writer working within a small division of a major manufacturing corporation, you and three colleagues witness a problem occurring within your corporation. The four of you believe that you know how to solve this problem, and a solution means that your team receives high praise from management and the

Susan's team meets with conflict during the brochure planning…

One major point hindering the progress of the "Mammography Services and Breast Cancer Prevention" brochure has been deciding the brochure's focus. Joseph, the public relations manager, believes that the brochure's focus should *only* identify and promote the mammography services that Crystal River provides. Gary, who maintains Crystal River's mammography equipment, believes that Crystal River's excellent record with federal standards and guidelines should be described *in detail* in the brochure as well. However, Jan, Dolores, and Norman—all medical staff who work closely with mammography and radiology patients—believe that the brochure should provide *only* guidelines and facts about mammography's benefits and its safety issues. The brochure should not, in Dolores' words, "sell mammograms to Crystal River women."

chance for promotions and raises. Therefore, your team submits a three-page, unsolicited proposal to upper-level management outlining your plan to solve the problem and describing how the solution benefits your organization.

Both of these instances, whether you respond to an RFP or to a problem within your own organization, allow you to use the proposal to help solve a problem. However, one important point to keep in mind is that proposals are not always successful. As a proposal writer, you must create the most persuasive proposal possible, since your proposal competes with other proposals written to "win the bid" for the same job.

Using an effectively written e-mail, Susan helps to facilitate teamwork…

Susan believes that with the brochure due at the printer in one month, planning must begin for this document immediately. To facilitate this planning, and to spearhead a compromise among her group regarding the content focus of the brochure, Susan writes the e-mail message shown in Figure 3.1 to all of her team members.

Figure 3.1 is an example of the type of important documents technical writers produce every day. While an e-mail message may not be considered a critical document in every situation, in the case of Susan's team and their brochure, it was an important document.

Figure 3.1 Excerpt from Susan's team e-mail

E-mail message contains a clear subject line that team members should understand.

Subject: WWP Team Brochure Planning
Cc:
Bcc:
Attached:

Hi team,

As we agreed in the meeting, the brochure planning needs to begin soon. We all seem to have good ideas about what information to place in the brochure. While it is only a double-sided, three-panel brochure, we still may be able to combine both focuses--mammography care and Crystal River mammography services--into it.

To generate some good ideas about how to do this, and to get a head start on our planning, I'd like to suggest that we each complete an "assignment" for Friday's meeting. If you have time between now and then, please draft an outline of the major points that you believe should be included in the brochure. Then, bring your outline to the meeting, so we can decide--as a team--how to proceed with the brochure.

The first paragraph contains information that gives background about the current situation. Providing this type of information first enables readers to better understand the unfamiliar information that follows.

The second paragraph contains information that suggests an action to be taken—in this case, each member preparing an outline of the brochure for the next meeting.

Reports

Reports are a type of document that nearly everyone in all sectors of business and industry relies on to help them get work done. Each type of report differs in the kind of information reported and the document's audience and use. Listed below are several common types of reports that are typically produced by technical writers in the workplace.

***Annual report*:** An annual report provides an overview and analysis of your organization's financial activities for a twelve-month period. Typically, annual reports are generated for an audience of stockholders, management, potential investors, or anyone else interested in the current fiscal health of your organization.

Issues of document design are carefully considered for annual reports, since readers usually equate an expensively designed and professional-looking annual report with a profitable and professional company.

Progress report: A progress report describes the progress that you, your project team, or your company is making on a particular task or project. Because of this broad definition, progress reports have a variety of audiences who may or may not work for your company. For example, supervisors use progress reports to track individual or team progress, while a client uses them to follow and track your team or company's work on a project. However, as a technical writer, you also may write progress reports to create an archive of the work you have performed.

Typically, progress reports contain a description of work performed (providing most detail for work performed since the last progress report), a description of current work activities, and a discussion of what tasks must be completed.

Recommendation report: A recommendation report identifies, describes, and then recommends the most suitable action, event, or item (such as a piece of equipment or a tool). For example, a typical recommendation report written for a construction contractor identifies and describes three types of commercial sprinkler systems, while it recommends the safest and most cost-efficient sprinkler for the contractor's specific needs.

Recommendation reports are written for an audience within or outside of your company, such as management, colleagues on a project, or clients. As a technical writer, you may write a recommendation report individually or collaboratively on a project team. Many times, recommendation reports are written after initial research—research conducted to identify the most suitable recommendation—has been performed.

Feasibility report: A feasibility report describes the consequences, or outcomes, of a particular decision or action. This type of report, like the recommendation report, helps its audience make decisions that are most beneficial in terms of cost, time, and personnel to their organization. Like the recommendation report, a feasibility report is preceded by a period of research or study in which a project team (or an individual) identifies and researches the possible outcomes of a decision or action.

Feasibility reports must carefully identify and describe not only the action that is being decided upon, but also the results or consequences of that action. In other words, the audience must clearly and completely understand the situation surrounding the decision.

Final report: A final report presents the results of a project and is submitted following the completion of the rest of the project work. Final reports accompany many types of projects, and nearly every organization in all types of business and industry creates and uses final reports.

A well-written final report provides an overview of the project's problem, a description of the major tasks performed to solve that project's problem, and the results of the work. Final reports may also contain overviews of the ways in which this work solved the project's problem.

The length, organization, and design of final reports are based largely on the length and complexity of the work that they are presenting. For example, if a project needed several months to complete, the final report will probably be lengthier and more involved than a report for a week-long project. Lengthier final reports need other elements—such as an executive summary, report abstract, or table of contents—to organize its content and guide readers through the document.

Trip report: A trip report identifies trip expenses and provides an overview of trip activities. Trips that you might write trip reports for include a meeting with clients in another city or country, a professional conference, or a tour of a factory or workplace.

Trip reports are useful not only for calculating your individual trip expenses, but also for describing the activities that you participated in during the trip that most impact your organization. The trip report is a handy way to explain how the information you gathered on your trip benefits your company.

Susan's team collaborates to produce an informational brochure…

Once Susan's team is in agreement about the focus of the brochure—largely because of the excellent discussion that resulted from Susan's e-mail—the team begins to write the brochure. It takes three meetings to complete the brochure, and Joseph sends the document to the printer shortly after the final editing meeting.

The brochure that the team eventually produces is shown in Figure 3.2. The team decided to make the brochure's focus a dual one: besides the one-panel title, two panels of the two-sided, three-panel brochure focus on the services that the Crystal River facility offers, and the remaining three panels provide statistics and information about mammography services and breast cancer detection and prevention.

Figure 3.2 "Crystal Community Hospital: Information Women Can Use About Breast Cancer Prevention and Mammography Services"

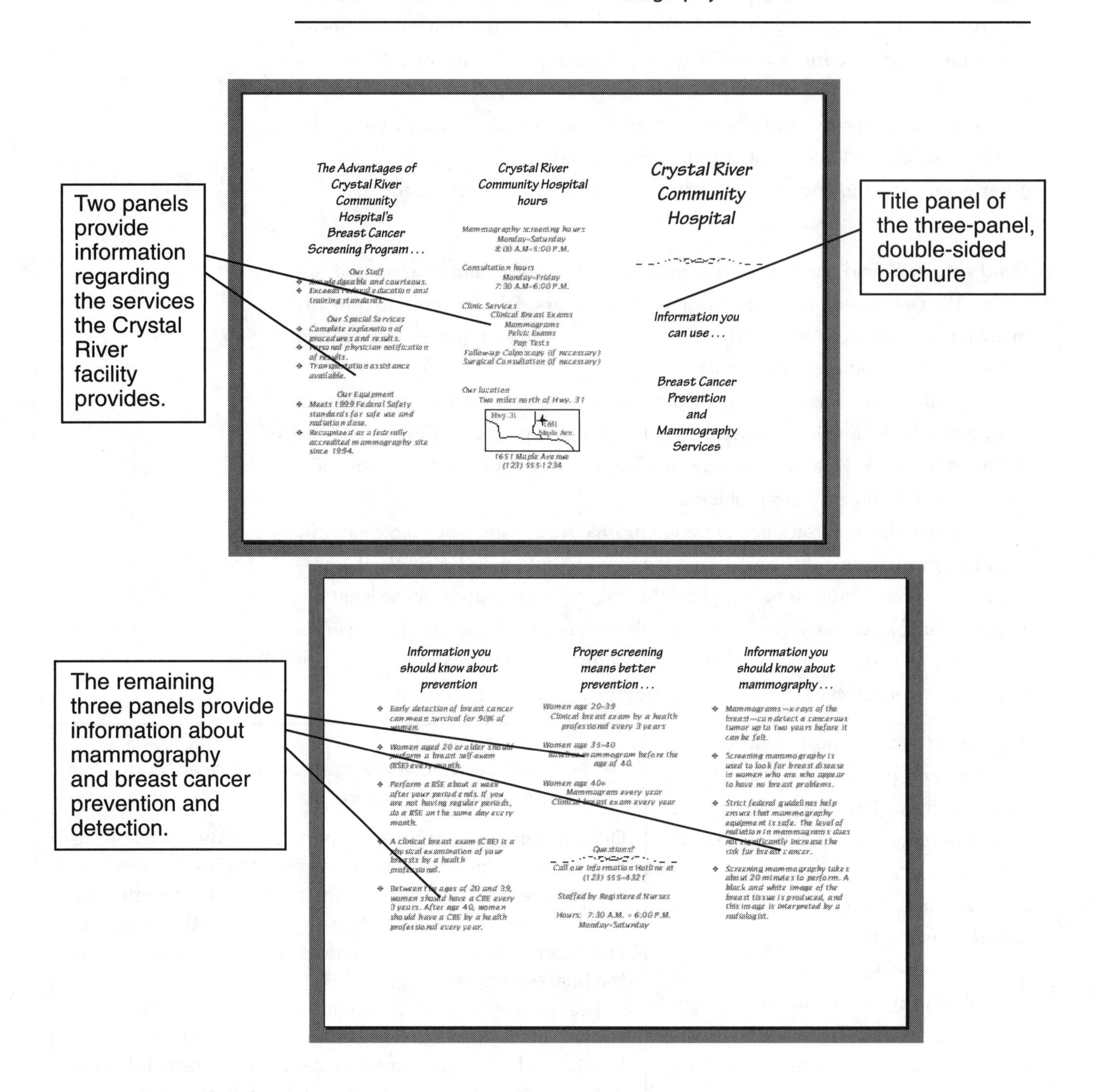

Oral presentations

Oral presentations are one of the most common projects in which technical writers are involved. As a technical writer, you may plan, prepare, and present an oral presentation individually or with a team of colleagues. Presentations are given to all types of audiences, including peers within your department or division, colleagues in other departments or at other work sites, clients, customers, or the general public.

Most businesses now require more from presentations, technologically speaking, than they did in the past. For example, as a technical writer, you must be familiar with at least one type of presentation software, such as PowerPoint, that will enable you to create visually interesting, well-organized, and professional-looking slide presentations. Most technical writers find presentations effective in connecting with a variety of audience members. Presentations enable writers to see and interact with their audiences, something that written documents usually do not allow.

Web sites

A company Web site is fast becoming a requirement for all businesses—small, large, and in nearly every sector. In more and more companies around the country, technical writers are being asked, because of their education and training with both textual and visual elements, to create and maintain company Web sites.

Besides a Web site, many organizations now use an intranet to help them do business more efficiently and effectively. An intranet is a series of electronic documents—from correspondence and reports to visual aids and archives—accessible only within the company by company employees. Your company's intranet—an electronic form of organizing, archiving, and communicating—may also be created and maintained by technical writers.

Response Exercise 3.1

Technical writers create many different types of products. Examine the list of documents and resources discussed in the preceding section. By visiting one or more local businesses, try to find examples of at least two or more of these types of documents. Or, visit the Web sites of two or more local organizations or businesses. Once you have obtained these materials and read through them, write down your responses to the following questions:

1. Write the name of the businesses where these materials were obtained. What products or services do these businesses offer? Have you done business with them in the past? If so, was it successful?

2. List the types of materials that you have obtained. For each type that you have listed, what qualities or elements led you to label it as such? For example, if labeled a document a type of "sales material," what qualities in that document made you label it, "sales material"?

What qualities characterize an effective product?

Technical writers attempt to include certain qualities within the documents and resources they produce. Qualities, such as an attention to communication context, effective prose style, completeness of content, purposeful organization, and appropriate design, are used by technical writers to help ensure the effectiveness of the materials they create.

The qualities we discuss here are very important for producing successful technical communication, and we continue to broaden this discussion throughout the following chapters.

Susan collaborates with Gary to produce an equipment specifications sheet...

Besides offering the Women's Wellness Program, Crystal River provides quality service to its community by complying with the changing federal regulations that standardize mammography equipment. One of Susan's jobs as a technical writer consultant for Crystal River is to work with Gary, a member of Susan's WWP team who works as a radiology department staff member maintaining the mammography equipment. Susan collaborates with Gary to help the radiology department meet federal mammography equipment standards. Both Gary and Susan decide to include some discussion of the importance of meeting federal equipment standards in their team's WWP lecture.

To comply with federal standards, Gary needs information about the current mammography equipment from the Crystal River region's federal equipment inspector. To request information, Gary works closely with Susan to develop a specifications sheet, or "specs," regarding the radiology department's current mammography equipment. These equipment specs, shown in Figure 3.3, are accompanied by a lengthy letter requesting information about the changing federal equipment standards. The federal equipment inspector uses this letter and the equipment specs to provide Gary with clear, complete, and accurate information about federal standards and Crystal River's mammography equipment.

Attention to communication context

In Chapter 5, we define in greater detail the communication context, and we discuss why it is important for successful technical communication. To define communication context for any document or resource, you must understand that document or resource's situation, purpose, and audience.

Situation refers to the set of events or circumstances that surround any given document. A document's deadline or a tight budget for printing and distribution are two circumstances that may contribute to a document's situation.

The *purpose* of a document refers to why the document is needed or what work the document helps your organization perform. For example, the purpose of a document may be to allow readers to operate equipment, learn more about the services their credit union offers, or take action against a specific event.

Audience refers not only to the readers of a document and those listening during a presentation, but also to those using a Web site. Each product has a specific target audience, such as clients, customers, or vendors. However, remember that most have more than one audience, and your document must meet the needs of these different types of readers.

As a technical writer, you must be acutely aware of the impact that communication context has on a document or resource. Understanding and attending to this context throughout the planning, drafting, revising, and editing stages of a document's writing process enables you to create a more effective, successful product.

Figure 3.3 Mammography equipment specifications sheet (excerpt)

Table that lists general specifications is clearly organized through a table title and number

Table 1. General Specifications ABC series Medical Film Digitizers

Feature	*Specifications*
Sensor	6,000 pixel linear CCD
Spatial Resolution	42 microns/pixel (600 dpi)
Gray Scale Resolution	16 bits/pixel (65,536 gray levels)
Input Media	Radiographic film up to 10 mm thickness
Autofeeder	Holds up to 25 films @ 7 mils
Film Sizes	2.5" wide x 4.5" long minimum 10" wide x any length maximum
Illumination	Cold Cathode flourescent
System Interface	16 bit high speed DMA Windows NT control software

Rather than using a paragraph format, a table with a concise listing of features and specifications provides complete and easily scannable information

Complete content

As a technical writer, you strive for the content of all of materials to be complete. That is, the content of any document should communicate all the necessary information to its readers. Therefore, not only must you understand the document's content, but also you must know your readers and their needs. Besides completeness of content, writers strive for content accuracy as well. In Chapter 12, we discuss the strategies that writers use to revise their documents for both completeness and accuracy.

As a technical writer, you will find that communicating content to your audience in a complete and accurate fashion is a challenging yet crucial aspect of your job. Frequently, writers who communicate with certain types of readers—particular clients, customers, or colleagues—find that creating a complete and accurate message for one type of reader is more challenging than for another. But ensuring that the message or content you communicate is complete and accurate leads to better results, and allows your organization to continue to successfully communicate with others.

Purposeful organization

The organization strategies we discuss in Chapter 6 are useful ways to organize or order information within most documents or resources. As a technical writer, you need to understand the most suitable method for organizing all of the major and minor points in your document. Readers respond to organization that is purposeful—whether that organization enables readers to understand a description of a process more easily or effectively persuades them to purchase a new piece of equipment—and all documents must be organized with the reader in mind.

In Chapter 5, we discuss ways to identify and use a document's communication context to help identify the most suitable and purposeful organization strategy. As a technical writer, you will rely on a variety of organizational strategies to effectively communicate your messages.

Susan's team reacts to the WWP lecture slides collaboration...

After Susan's WWP team completes the informational brochure for the WWP participants, they must begin organizing the WWP lecture, or oral presentation, about mammography services and breast cancer prevention.

Jan, Dolores, and Norman use their medical expertise to create most of the content for the lecture, "Mammography Services and Breast Cancer Prevention." Susan's communication and visual-design knowledge help the team to produce an aesthetically appealing and well-organized slide presentation. Their team also benefits from Joseph's public relations knowhow to help sell the WWP and the hospital's mammography services. Gary's expertise as a medical physicist is beneficial as he includes basic information about Crystal River's mammography equipment, and he notes that the equipment meets federal standards (standards met, in part, by Gary and Susan's collaboration on the equipment specifications document). The team agrees that their collaboratively planned presentation provided greater detail in a more effective manner than any individually planned presentation could have.

Effective prose style

In most documents and other resources, an effective prose style is built on three elements: clarity, conciseness, and courtesy. Of the three, clarity is perhaps most important. *Clarity* refers to the precision and focus of the message that you are communicating. Clarity is an element that enables an audience to understand your message without further explanation.

A *concisely* written document completely conveys its message using the minimum number of words. If a document is concise, its audience can read and understand its message quickly and efficiently. *Courtesy* is an important element in all communication (not just correspondence), since a discourteous message may be off-putting enough to the readers that it hinders the message of even the most clear and concise document.

Creating material that exhibits a clear, concise, and courteous prose style can be a very difficult task. However, as a technical writer, your collaboration with colleagues and editors makes revising for this prose style easier.

Appropriate design

Document and page design are important aspects of any type of technical resource. Frequently, as a technical writer, you will spend as much time planning and creating the design of your document as you do drafting and revising that document.

An appropriately designed document takes into consideration that document's communication context. The document's situation, purpose, and audience reveal a great deal about how a document should look on the page. Chapter 9 discusses how communication impacts design issues and what strategies technical writers use to produce a usable and aesthetically appealing document.

When a document or resource is thoughtfully designed, the content or message is communicated more clearly. As a technical writer, one of the challenges you face with every project is creating a design that supports and complements the message you communicate.

Response Exercise 3.2

Technical writers attempt to include certain qualities within the materials they produce. Qualities, such as an attention to communication context, effective prose style, completeness in terms of content, purposeful organization, and appropriate design, are used by technical writers to help ensure the effectiveness of the documents and resources they create.

To learn more about how these qualities are reflected in different types of materials, obtain two or more types of documents. These could be Web sites, sales materials, reports, proposals, or correspondence. After reading the material, write down your responses to the following questions about each item that you have obtained:

1. After studying the item, what information do you have about its *communication context*? What is its situation, purpose, and audience?
2. After having read the item, does its *content* seem complete and accurate? If so, identify those details that were particularly helpful. If not, what other information would you include?
3. Sketch an outline of the order of points presented in the item. Does its *organization* seem purposeful? If so, identify one or two elements that contribute to its logic. If not, how would you revise this organization?
4. After having determined the communication context of the item, does its *prose style* seem useful and appropriate? Identify at least two passages or sentences that are either effective or need revision.
5. Would you characterize the *design* of your item as complementary to its message and purpose? Describe why or why not.

Chapter 3 Task Exercise

This Task Exercise asks you to conduct the third of a four-part interview with a technical writer. Identifying and describing the work performed by this writer at his or her workplace is the topic of this interview. (Task Exercises in Chapters 1, 2, and 4 describe the remaining three interview parts.)

Interview a technical writer about one document or resource that he or she has recently produced. This could be a Web site, letter, report, proposal, type of sales material, or any product created by the writer. Using the interviewee responses, create a genealogy chart that traces the "life" of the document or resource. You may include details about how and why it was initially created, how it was revised and by whom, and what kinds of work it helped perform at the writer's organization.

Susan's team completes the WWP lecture slides...

While all of Susan's WWP team members are present for the March lecture, her team decides that Jan, a registered nurse, and Dolores, a medical imaging technician, should actually present the lecture and lead the question-and-answer period that follows, since women coming to Crystal River for mammograms would likely encounter one of these two women.

The PowerPoint slide presentation, entitled "Mammography Services and Breast Cancer Prevention," was the most time intensive and challenging resource ever produced by the group (Figure 3.4). However, Susan believes that it is one of the best presentations she has ever helped to create, and for that reason, the time and energy were well worth it.

You may organize and design your genealogy chart in any way—as long as it communicates your points about the life of the document or resource in a usable and appealing way. For example, you may use poster board to create a large and

Figure 3.4 PowerPoint slides 2 and 3 from "Mammography Services and Breast Cancer Prevention"

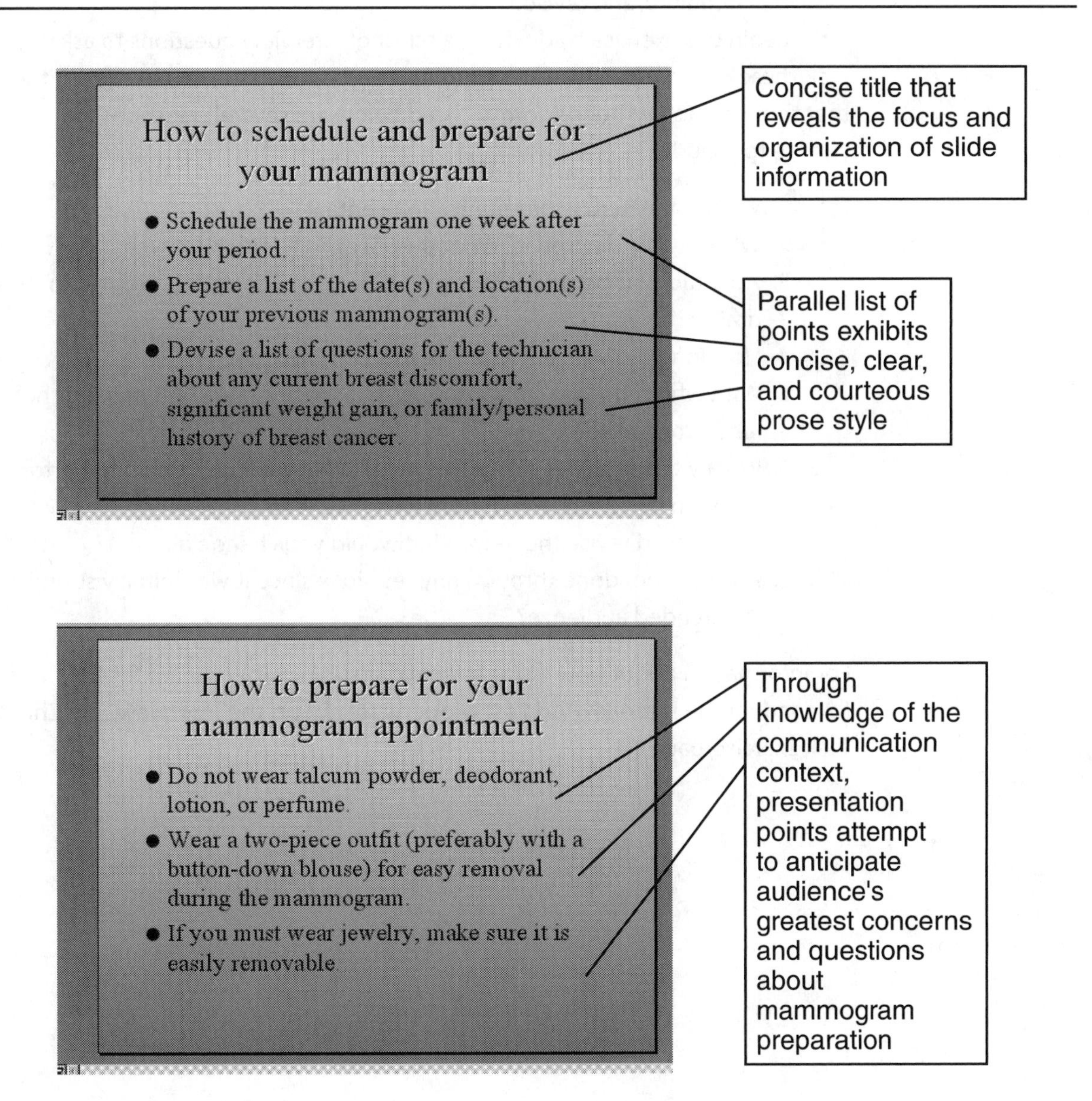

colorful chart, or you may use a software application to create a chart that can be easily saved, edited, and revised.

Begin this exercise by devising a script of interview questions to ask your subject. These questions must attempt to find out as much as possible about the work that this technical writer performs. Listed below are several questions that you may choose to include.

1. What is the item's communication context?
2. Who initially devised it? What purpose did it initially serve?
3. How did you begin planning it? What considerations went into its early drafts?
4. How long did you plan/draft/revise/edit this item?
5. With whom do you collaborate? What specific contributions did these collaborators make?
6. What was the end result of this item? What work did it help to perform?
7. What do you believe was most successful about it?
8. If you could revise the item, what would you change about it?
9. Has the item gone through any revisions since it was initially submitted to its intended audience?

For information about how to (a) select your subject, (b) prepare for the interview, (c) conduct the interview, and (d) properly follow-up the interview, see Chapter 1 Task Exercise, page 13.

Chapter 4

How Do You Become a Technical Writer?

To help you understand more about becoming a technical writer, you should know what interests, experiences, and education may help you to better qualify for a career in this field.

As a technical writer, one of your main responsibilities is to help professionals in a variety of occupations communicate effectively and efficiently. This, in turn, enables those professionals and their organizations to become more productive, efficient, and successful.

As trained professionals who work in a variety of organizations, technical writers both produce written, oral, and electronic communication, and help other professionals to communicate more effectively. Typically, those who wish to become technical writers already have a genuine interest in communication (its processes and benefits) along with an interest in a related business, technical, or scientific field. Aspiring technical writers generally obtain a college degree—whether it is a two-year, four-year, or graduate degree.

In this chapter, we discuss how you can decide whether technical writing is the right career for you. To help you make this important decision, we identify how your interests, experiences, and attitude toward education help to qualify you as a potential technical writer. We also discuss what you must consider when dedicating yourself to a degree program. Then, we identify several questions to ask about education programs, and we provide you with a sample of currently available programs.

To understand how to begin a career in the technical communication field, you should know the answers to the following questions:

- What *qualities* best qualify you for a technical writing career?
- What *program of education* is right for you?
- What *types of programs* are available?

Understanding the ways in which many people prepare for a career in this field allows you to decide whether or not technical writing is the job for you. Remember, while the information we present in this chapter is not exhaustive, it does provide you with facts to think about regarding your important career decision.

What qualities best qualify you for a technical writing career?

If you are interested in a career in technical communication—whether you are a recent high school graduate, a college student, or a professional interested in making a career change—you can better understand this field if you know what interests, experiences, and attitudes toward education best qualify you for a career as a technical writer. While there are no hard-and-fast prerequisites for those seeking a career in technical communication, understanding what qualities may make you more or less suited to this type of occupation is important.

In this section, we identify qualities that those interested in a career in technical communication typically possess. For example, communication and collaboration are typical *interests* among those who pursue a career in this area. Prior work in a related field is an *experience* that characterizes many potential technical writers. A belief in the usefulness of pursuing a degree and other *positive attitudes toward education* are also common among those interested in this field.

Lawrence's story…

Lawrence Dawson works as an assistant manager at a local restaurant chain. For the last three years, he has worked full time to help support his family—his wife, Dominique, and their two-year-old son. Currently, Lawrence's main responsibilities include managing a staff of sixteen, completing food purchase orders, and devising the staff's shift schedule. Managing his staff requires effective communication, and Lawrence rises to the challenge by using primarily oral communication, as well as an occasional memo. Lawrence enjoys communicating on the job. Lately, he has been thinking of ways to parlay this interest in communication into a satisfying new career.

Interests

As someone who is considering a career in technical writing, your interests should be broad because the topics you write about will be widely varied. Whatever your primary interests may be, they should relate in some way to written and oral communication, collaboration, and technology.

An interest in communication includes an enthusiasm not only for writing and orally presenting information, but also for using a variety of forms of technology to facilitate communication. An interest in collaboration includes a willingness to work with others to complete a task or a project and the ability to give and take constructive criticism in a professional manner. An interest in technology includes not only a general interest in the uses of technology in the workplace, but also a healthy perspective regarding both the benefits and drawbacks of using technology to communicate. To better understand the technology that technical writers use, refer to the basic software applications listed in Table 4.1.

Table 4.1 Software applications technical writers use

Basic operating systems and applications	*Intermediate applications*	*Other operating systems*
• Windows 98, Windows NT • Office 98, MS Word	• PageMaker, PowerPoint • FrameMaker, Publisher • PhotoShop, Excel • Access, Visio	• UNIX • Macintosh

In its brochure, *Careers in Technical Communication,* the Society for Technical Communication (STC) notes that to become a technical writer, "you must convince a prospective employer that you have an aptitude for technology and that you can communicate technical information in a manner that various audiences can easily understand."

Along these lines, those pursuing careers in technical writing often develop an interest in a corresponding field. Fields that complement technical writing include those in business, technical, and scientific areas, such as business management, management information systems, computer technology, Web design, engineering, biology, agriculture, chemistry, or botany.

Experiences

While many successful technical writers have moved directly from high school to a degree program to the workplace, it is beneficial to have varied life experience to complement your career in this field. Your work experience in another organization, your volunteerism, or your entrepreneurial work may positively influence your career in technical writing. For example, if you have a great deal of volunteer experience with a community charity organization prior to becoming a technical writer, you may choose to work for a non-profit agency after you have received your degree in technical communication.

Technical writers often come from backgrounds in other related careers. Those who "change horses" to become technical writers might come from careers in business, industry, or the sciences. Individuals with prior work experience are often highly recruited by prospective employers. Employers understand that writers who bring a variety of experiences with them to their current positions may offer important insights about the workplace in general and may be more adept at communicating with a variety of audiences.

Other valuable life experiences might include volunteer work in which you were responsible for a great deal of your organization's communication, whether at a non-profit organization, church, or school. Another valuable professional experience includes entrepreneurial involvement that you might have had with your own or someone else's company. University admissions personnel and employers alike look for candidates with these types of work or work-related experiences.

Lawrence begins to analyze his career options...

Using Dominique's suggestion, Lawrence decides to analyze career options in the communication field. Lawrence's interests include professional writing, communicating with colleagues, and taking the leadership role in team situations. While his computer experience is rather limited, Lawrence is very interested in computer technology—particularly in the World Wide Web and e-commerce businesses. After high school, Lawrence completed a two-year business management degree at a local community college.

Lawrence decides to visit the public library, examine the career guides and college handbooks that are available, and further analyze his career options.

Education

A quality that anyone interested in pursuing a career in technical communication should possess is a willingness to commit to an education program, whether it is a two-year, four-year, or graduate degree program. This commitment requires a dedication that involves time, money, and a positive attitude.

Depending on your program, you may spend anywhere from two to seven or more years pursuing a degree. If you seek your degree as a part-time student, this timeline may be even longer. Choosing a career in technical communication means investing the time to *complete* a degree. Remember, though, time invested now is time that you do not need to invest later.

Regardless of the type of degree program you choose, you spend money pursuing that degree. While creating a budget that specifies tuition, books, room, board, meals, and other essential expenses is a necessity for most technical communication students, assistantships, financial aid—in the form of loans, grants, and work study—and even paid internships can all help to pay for your education.

When you enroll in a program, you should have a positive attitude toward that program and toward learning in general. A positive attitude means that you are ready to accept the responsibilities of learning and willing to gain the most from your education. Your positive attitude may mean the difference between simply receiving a degree and receiving a diverse, worthwhile, and well-rounded education.

Response Exercise 4.1

In the preceding section, we identified qualities that those interested in a career in technical communication may typically possess, including interests, experiences, and attitudes toward education. To better decide what types of qualities you possess, write down your responses to the following questions:

1. Using the information in this chapter, describe how you possess the three types of *qualities*—interests, experiences, attitude toward education—that help to qualify you for a technical writing career.
2. Using complete sentences, identify and describe your *interests*. How are these compatible with a technical writing career?
3. Identify and describe your *experiences*. These may be any type of professionally oriented experience. How do these prepare you for a technical writing career?
4. Identify and describe your *attitude toward education*. In what ways are you willing to invest time, money, and learning attitude toward your degree?

What program of education is right for you?

Three broad categories of technical communication education programs exist—two-year, four-year, and graduate degree programs. While STC did not report on two-year degree salaries in its 1998 *Technical Communicator Salary Survey*, it did report that the mean salary—based on education level—was $47,300 for those with bachelor's degrees, $49,460 for those with master's degrees, and $59, 480 for those with doctorates.

To better determine what type of education program is right for you, think about the follow-

Lawrence thinks about a career in technical communication…

After his visit to the library, Lawrence pores over the materials he checked out and the notes that he took regarding potential careers. Lawrence discovers that communication is a skill that is important to many different types of fields. However, he is intrigued by a career in technical writing, as it seems to be oriented toward both communication and technology. To become a technical writer, Lawrence's degree in business management is not enough. He and Dominique begin to discuss the option of Lawrence returning to school for a four-year degree in technical communication.

ing three factors—time, money, and program mission. While in no particular order, these three factors most readily influence the type of program that you choose.

Time

The amount of time that can you afford to devote to a degree program helps to determine what type of program you choose. For many, a four-year degree is possible, while for others a two-year degree is most feasible. Remember that the time you devote to your education is an investment. However, only you can decide how the time factor influences the type of degree program you choose. To help you determine this, ask yourself the following questions about time:

Questions to ask yourself about *time*

- What type of time investment can I make? What factors, such as family responsibilities, financial obligations, or personal goals impact this proposed investment?
- To what degree are my family members willing to support me—personally, financially, or otherwise—during the time I plan to pursue my degree?
- What alternative or contingency plan do I have if I am unable to complete my program?
- In what ways—social, economic, and personal—will the time investment I make pay off in the future?
- In what ways—social, economic, and personal—will the time investment I make be detrimental to me in the future?

Remember, in many cases you should consider the time factor in a broad context. For example, you may have to consider what effects this factor has on those family members or friends who may be most affected by your decision.

Money

The amount of time that you can afford to devote to a degree program is directly related to the amount of money that you can invest in a degree program. For many, a four-year degree is a goal that you and your family have considered and planned for many years. In a financial sense, a four-year or even a graduate degree program may be more readily attainable in these cases.

Financial aid—such as work study, loans, assistantships, scholarships, and grants—can benefit nearly anyone interested in pursuing a degree. Those who must return to school in order to make a career change, along with other more "non-traditional" students may be eligible for additional financial aid. You should find out as much information as possible about financial aid before committing to a specific degree program.

Like the time investment, the financial investment that you make must be carefully considered, planned, and implemented. Listed below are several questions that you should ask yourself about this financial investment:

Questions to ask yourself about the *financial investment*

- What yearly budget will I need to complete the program of my choice? What specific school expenses will I have—such as tuition and books—and what personal or living expenses will I have—rent, food, transportation, etc.?
- What funding sources—personal, family, or institutional—can I depend on? What specific investment will they make?
- How will the financial investment I make now pay off in the future?
- How will the financial investment I make now be harmful to me in the future?

Program mission

Simply choosing to invest time and money in a degree program is an important decision. But to make your investment truly worthwhile, you should understand each program's mission. Basically, a program mission is the goal that each program wants to help you attain.

Every program mission should be fully supported through that program's curriculum, resources, faculty, and career placement strategies. Table 4.2 identifies and explains each of these four categories, and it provides you with questions to ask about each program.

Lawrence and Dominique think about education programs...

Lawrence and Dominique think carefully about the plan for Lawrence to return to college. Both he and Dominique discuss the factors of time and money. They agree that any time investment that is made now will not need to be made in the future. Also, Lawrence points out that a technical communication degree is a degree in a field that is "open"—that is, job opportunities are available in a number of different sectors of business and industry. The financial investment is something Lawrence and Dominique discuss thoroughly. Lawrence decides to examine two programs in the area, while he pays specific attention to the financial aid that each offers.

Table 4.2 Program mission criteria

Criteria definitions	*Questions to ask*
Curriculum	
• Offerings • Variety • Focus • Balance	• Does the curriculum offer courses in *practice* (analyzing, writing, designing, editing documents), *theory* (theoretical positions regarding why to choose specific practices), *teaching* (college classroom, corporate training seminars), and *research* (methods for conducting workplace, academic research)? • How many courses are offered each semester/quarter? How frequently, per academic year, is each course offered? • Is there a balance of theory, practice, teaching, and methods courses offered each semester/quarter?
Resources	
• Computer • Library • Technology • Tutoring	• How many university- and department-wide computer facilities are available for student use? What types of software are available for student use? • How many volumes does the university library own? How many subscriptions in professional communication-related fields does the library carry? What are the library's multimedia capabilities? • What other technological resources—providing students with access to a variety of multimedia equipment—does the university have available? • What university- or department-wide tutorial services—from writing, class-specific, to personal support—are provided?
Faculty	
• Number • Background • Connections	• How many full-time faculty teach professional communication courses? How many part-time faculty teach professional communication courses? • What are the education/expertise/training/interests of each full-time faculty? What are their connections with industry?
Career placement	
• Internship/Co-op • Job strategies	• What internship or cooperative education experiences are offered? If offered, do experiences include a variety of businesses and industries? • What university- or department-wide seminars/conferences/classes are offered in job strategies, such as résumé and interview preparation? How frequently are career fairs held?

To help you gauge the suitability of your program choices, you may ask these questions to the program's advisor or a university contact person. A wealth of information is available through the World Wide Web, since many universities have Web sites containing extensive amounts of information for prospective students. Many times, questions such as those posed in Table 4.2 can be answered by visiting the university or department Web site, or by contacting the university via e-mail.

In the next section we introduce you to a selection of degree programs—two graduate programs, two four-year programs, and one two-year program. To obtain more information about technical communication degree programs in your area, be sure to visit your local library for more information or access Web sites of colleges and universities in your area.

Response Exercise 4.2

In the preceding section, we identified three factors that you must consider when choosing an education program. These factors include time, money, and program mission. To better decide how these factors relate to your program of choice, write down your responses to the following questions:

1. Using the information in this chapter, define the three main factors—time, money, and program mission—that should impact how you choose an education program.
2. Using complete sentences, identify and describe how the *time* factor impacts your program choice. For example, is your choice of a two-year program a direct result of the time factor? If so, why?
3. Identify and describe how the *financial investment* factor impacts your program choice. Create a budget for the first year at the program of your choice—remember to include estimated financial aid and all necessary expenses.
4. Identify at least two programs of interest to you. Respond to the questions in Table 4.2 regarding their *program missions*. How does each program rate? Based on these criteria, what program would you choose? Why?

Lawrence investigates both education programs...

Lawrence examines two four-year technical communication programs in his area. He accesses information about them at the library—via the Internet—and prints out the necessary program facts. Lawrence evaluates each program using four factors: curriculum, resources, faculty, and career placement strategies. Shorewood Technical Institute—with its four-year degree and varied assortment of courses—seems to be the program of choice. Shorewood also supports more on-campus computer labs, and services a small but useful multimedia facility. Lawrence discovers that Shorewood's career placement center offers a variety of services—from job listings and job interview seminars to career counseling.

What types of programs are available?

The programs described below—two graduate, two four-year, and one two-year—represent a sampling of available programs in technical communication. These programs were selected because of their geographical diversity, and the subtle differences of their program missions.

The information provided about each program was gathered during summer 1999, and, when available, information was obtained via the Internet (each university or college's URL—Web site address—is listed in the subheading). The following information is meant to introduce you to the variety of programs being offered, and it is not meant as a complete description of each program. If one of these programs piques your interest, refer to that program's Web site or contact them directly for more information.

Bellevue Community College <www.bcc.edu>

Bellevue Community College (BCC), located in Bellevue, Washington, offers a variety of courses to its over 18,400 students. Courses that focus on vocational subjects, job retraining, and basic educational enrichment make BCC a popular education option. For the last three years, BCC has had a 77% average job placement rate in its occupational programs.

BCC offers several communication-intensive computer and multimedia programs. One program, Information Technology Programming, educates students to be entry-level programmer/analysts. The Media Communication and Technology program—either a one-year certificate or a two-year degree—enables students to learn the production and application of computer, video, animation, and Web design technologies. Students use these skills to create, use, and maintain media-based materials for business, industry, and education.

Florida Institute of Technology <www.fit.edu>

Florida Institute of Technology (FIT), located in Melbourne, Florida, is an independent, technological university with over 4,200 students. The university houses six academic colleges, including engineering, aeronautics, and science and liberal arts.

The communication department, housed in the Science and Liberal Arts College, offers a BS degree in Communication, with a specialization in either Business and Marketing Communication or Scientific and Technical Communication. Among the strengths of its program, FIT cites its varied curriculum and strong university admissions policy.

A BS degree in Communication requires 120 total credit hours to graduate. Besides the humanities and sciences requirements, students are required to take several communication courses, which include courses in rhetoric, composition, professional writing, visual communication, layout and design, editing, mass communication, research, and professional presentations. Seniors with good grades—a 3.25 GPA or over—may apply for internships.

Carnegie Mellon University <www.cmu.edu>

Carnegie Mellon University (CMU), located in Pittsburgh, Pennsylvania, is a national research university with nearly 7,500 students. The university houses a broad range of academic departments, including a nationally recognized program in writing and rhetoric.

The English department offers ten different degrees—among them, a BS in Technical Writing, a BA and an MA in Professional Writing, and a Ph.D. in Rhetoric. Among the strengths of its program, CMU cites its interdisciplinary resources and distinguished faculty.

Students in the BS degree program in Technical Writing—after having taken general education courses—choose from writing courses emphasizing exposition, rhetorical studies, writing for specific audiences, and writing in the different forms of business and technical communication. This program is highly structured, and students follow a rigorous course of study that includes courses from the humanities, natural sciences, technology, and social sciences. Students who maintain a B average in writing courses usually take an internship during their senior year. The internship provides approximately 100 hours of professional experience.

The MA in Professional Writing requires 37 credit hours beyond the BA to graduate. Students choose from courses that include rhetoric, linguistics, style, research methods, professional writing, computer applications, desktop publishing, visual design, and organizational management. To complement their core curriculum, students must also either self-define or select a prescribed concentration: business, research, rhetorical theory, or a technical area. Students typically take a professional writing internship during the summer between their second and third semesters.

Iowa State University <www.iastate.edu>

Iowa State University (ISU), located in Ames, Iowa, is a research university with nearly 25,400 students. The university has nine academic colleges, including agriculture, business, education, and liberal arts and sciences.

The English department at ISU, housed in the College of Liberal Arts and Sciences, offers an MA with a concentration in Rhetoric, Composition, and Professional Communication, as well as a Ph.D. in Rhetoric and Professional Communication. Among the strengths of its program, ISU cites its active faculty, and the broad and varied curriculum it offers.

ISU's Ph.D. degree in Rhetoric and Professional Communication requires 72 total credit hours beyond the BA to graduate. Students choose from courses in rhetoric, composition, linguistics, rhetoric of science, professional writing, research methods, editing, Web design, visual communication, and teaching writing and communication. Students must select a related field—such as statistics, design, or the history of technology—as an appropriate extension of their core curriculum. Internships are available, but not required, for all graduate students.

New Mexico State University <www.nmsu.edu>

New Mexico State University (NMSU), located in Las Cruces, New Mexico, is a research university with a 15,400-student enrollment at its main campus. NMSU dedicates itself to research, service, and teaching, while enrolling a 44% minority population.

The English department offers degrees at all levels, including an MA with an emphasis in Technical and Professional Communication, as well as a Ph.D. in Rhetoric and Professional Communication. NMSU strives to accept students from varied backgrounds with strong academic and professional experience.

NMSU's Ph.D. degree in Rhetoric and Professional Communication requires 78 credit hours past the BA to graduate. Students choose from courses that include rhetoric, composition, research methods, professional writing and communication, and computer applications. Students are also required to enroll in at least six credit hours in a professional writing internship.

Lawrence and Dominique decide to invest in a technical communication program…

Lawrence and Dominique decide that Lawrence should enroll in the fall semester of Shorewood Technical Institute. They have spent several weeks talking with Shorewood's university advisors, members of the technical communication faculty, and a financial aid counselor. After these discussions, both Lawrence and Dominique believe that a degree in technical communication is a wise investment.

Response Exercise 4.3

In the preceding section, we identified and described several education programs, including two-year, four-year, and graduate degree programs, and a variety of program missions.

Access information on at least two programs of your choice that were not introduced and described in this section. (See Appendix A for ideas.) Using what you know about selecting a suitable program, write down your responses to the following questions:

1. What types of degree programs are offered? Is the selection of technical communication courses broad and varied?
2. What types of financial aid are offered by the university? The department?
3. Briefly identify, in two or three short paragraphs, the program's mission. Use the questions in Table 4.2 to help you define this.
4. Based on the program mission and your individual criteria (degree offered, financial aid, location, etc.), select your program of choice. What are your reasons for this selection?

To help you decide whether or not you wish to pursue a career in the field of technical communication, you should know what qualities—including your interests and experiences—best qualify you for a technical writing career. You should also consider both what type of education program is right for you and what types of programs are currently available.

Chapter 4 Task Exercise

This Task Exercise asks you to conduct the fourth of a four-part interview with a technical writer. Identifying and describing the ways in which this technical writer began in the technical communication field is the topic of this interview. (Task Exercises in Chapters 1, 2, and 3 describe the remaining three interview parts.)

Interview a technical writer about his or her interests, experiences, and education background—prior to becoming a technical writer. Using the interview responses, create a two-page "letter of reference" that describes the qualifications of this technical writer. Imagine that you are writing this letter to the technical writer's prospective employer. You may include as many details about the technical writer's interests, experiences, and education as you believe necessary.

You may organize your letter in any way—as long as it communicates your points about the professional "life" of the writer in an understandable and appealing way. For example, you may choose to describe his or her qualifications chronologically, by topic, or in another manner.

Begin this exercise by devising a script of interview questions to ask your subject. These questions must attempt to find out as much as possible about this technical writer's background and qualifications. Listed below are several questions that you may choose to include:

1. What interests led you to technical communication as a career field?
2. What experiences impacted your technical writing career?
3. What education and training do you have?
4. How have you used or benefited from these interests/experiences/ education?
5. What advice would you give someone interested in a career in technical communication?

For information about how to (a) select your subject, (b) prepare for the interview, (c) conduct the interview, and (d) properly follow-up the interview, see Chapter 1 Task Exercise on page 13.

Part II

Planning Effective Documents

Chapter 5

The What, Why, and Who of Technical Communication

To become a good technical writer, you need to identify the communication context—the situation, purpose, and audience—of every document.

In order to plan a piece of technical writing, you must begin by identifying three elements that make up the document's *communication context.* Understanding the communication context means identifying and defining these three elements:

- The *situation* that necessitated that document
- The *purpose* of the document
- The *audience* for the document

Knowing these three elements helps you move forward in the writing process. Once you understand the communication context, you can begin to organize and draft the document. Before the document is complete, you must decide what information to include, how to support and discuss that information clearly and concisely, and what your document will look like.

Understanding the communication context is the first step in writing a successful document. In this chapter, you will learn how to define these three important parts of the communication context—situation, purpose, and audience—and be able to discuss the strategies technical writers use to identify them.

What's the situation?

Once you are assigned a document to write, you try to understand as much as you can about that document's situation. In the workplace, documents are always linked to actual events, tasks, circumstances, and people. In order to write an effective document, one that is appropriate to its actual place and time, you must understand as much as you can about the writing situation. To do this, ask yourself questions about *time*, the possibility of writing with *collaborators*, and any previous *knowledge* that you have about the document's topic. Also, many writers keep handwritten or electronic notes about questions such as these, jotting down their preliminary ideas and impressions. This is a good habit to get into.

Connie's story…

Connie Smith, a technical writer at a large computer manufacturing firm, Compu-Trends, has been asked by her supervisor to create a brochure that informs customers about Compu-Trends' new flat-screen computer monitor.

Connie's supervisor tells her that the brochure will be included in the newsletter that customers receive at the end of the month. The supervisor gives Connie three pages of a hardware description that the engineering department has drafted describing the features of the new monitor.

Time

Time is important to consider in every writing situation. Typically, you must ask yourself these important questions about time, including questions about project schedule and deadline information:

Questions writers ask themselves about *time*

- What's the deadline for this writing project?
- How many months, weeks, or days do I have to write this document?
- How can I manage my time to meet this writing deadline?

You must be aware of project deadlines and the amount of time you have to complete a document well before you begin to write it. Once you understand project deadlines, you can set personal writing goals for yourself. For example, knowing that the product sales brochure must be completed in three weeks allows Connie to plan her writing schedule for that document over the course of those three weeks.

In order to write quality documents on schedule, you must be skilled at time management. However, not only must you be aware of your schedule, but, at times, you must be aware of others' schedules, because often you work in a team to write the document. That writing team may consist of one or more of your co-workers. Therefore, the next element of the writing situation you need to identify is whether or not a writing project is *collaborative.*

Connie considers time…

Connie's first task is to identify the communication situation. First, Connie considers time. She has only three weeks until the newsletter is sent out to her company's customers—Connie has little time to waste.

Connie decides to make this document her top priority in the coming weeks. She does "backwards planning." This means that she begins her schedule with her deadline and then plans backwards to fit in all of this project's tasks.

Collaborators

While you write many documents on your own, sometimes you collaborate with others. For example, if you write an instruction manual for a new product, you may collaborate with the engineers who helped to design that product and with the technicians who helped to build it. Also, you may collaborate with members of the design department to plan and discuss the manual's design or packaging requirements.

Questions writers ask themselves about *collaborators*

- Will I be working alone, or will I be writing within a team?
- If I am writing on a team, have I worked with members of this group before? What do I know about their work habits?
- What tasks is each group member responsible for completing?
- How will I communicate with my collaborators to produce a quality document?

Writing collaboratively is often a time-consuming and complicated process; however, the benefits of writing a team-produced document are many. In Chapter 10 we discuss not only the benefits of a team-produced document but also the strategies technical writers use to work successfully within teams.

Whether you write a document individually or collaboratively, you must think about one more element of the writing situation: the previous *knowledge* that you have about the document and its topic.

Response Exercise 5.1

Time and the possibility of *collaborators* are two elements that technical writers consider when they begin to identify the writing situation of a document.

Write down your responses to the following questions. Identify when you collaborated with others to complete one task: on the job, at school, or with members of your family. What was the task? Who were the collaborators?

1. What concerns with time did you have during the course of the task? Did you have to revise your schedule to fit the schedules of others?
2. Could you have performed this task alone? Why or why not?
3. If you had performed the task alone, would the outcome have been the same?
4. What were problems you encountered when working with your collaborators?
5. What were the benefits of working with your collaborators?

Knowledge

Your previous knowledge of the document will vary with the topic and your experience with it. The more technical communication projects you have done, the more topics you will be familiar with.

Identifying your prior knowledge about a topic lets you know what new information you may need to find in order to begin writing. You can ask yourself several questions to discover your previous knowledge about the document and its topic:

Connie considers previous knowledge...

Connie understands that the brochure she is preparing will not be written collaboratively; the brochure is her individual project.

Connie uses the three-page hardware description that her supervisor gave her to familiarize herself with the new product. After reading the hardware description, Connie still has questions about the product. Namely, what screen sizes are available, and what is the adaptability of the monitor to her customers' current systems?

Questions writers ask themselves about *knowledge*

- What previous knowledge do I have about the topic of this document?
- What knowledge do I need to begin drafting the document?
- Where and how will I access this knowledge?

If you *do* need more information, often you can find it within your company. Your company may have similar documents on file, data on file that supplement the contents of the document, or information in a database that will help you find out more about the topic.

Connie accesses more information…

Connie decides to ask Susanne, an engineer in the hardware design department, her questions about the new monitor. Connie has worked with Susanne before and knows that she usually responds promptly.

Instead of attempting to track down Susanne in person, Connie sends her an e-mail message requesting the necessary information.

Frequently, though, your co-workers are your most valuable information resource—even if you are not working on a collaborative or team project. To get more information about a project, you may request specific types of information—in person, during a telephone conversation, or via e-mail—from your co-workers. Your co-workers may tell you more about the project or about how the document is useful to that project.

Time, collaborators, and knowledge are the elements that you must identify in order to understand a specific writing situation. Next, you must know the document's *purpose* and *audience* to have a full understanding of the communication context.

What's the purpose?

After determining the document's situation, you must identify the purpose for writing that document. To understand this, you must ask yourself questions about the document's purpose before you begin to write it:

Questions writers ask themselves about *purpose*

- What is the objective or goal of the document?
- What work will this document do?
- Why is this document important?

Understanding the purpose of the document helps you to move forward in the writing process. Listed below are some typical purposes and situations of documents that you may encounter when you do on-the-job writing:

- ***To inform***
  As a technical writer for an insurance agency, you are asked to compose a short letter informing your firm's customers about a policy change in automobile insurance.

- ***To persuade***
  As a writer for a retail electronics outlet, you are asked to write and design a product sales brochure persuading customers to purchase your company's new cellular phone.

- ***To recommend***
  As a writer for a large manufacturing plant, you are asked to help write a report recommending a solution to your plant's current inefficiency problem.

- ***To instruct***
  As a writer for a computer software design company, you are asked to write and design a manual instructing customers about the use of your company's newest software package.

- ***To inquire***
  As a writer at a large, regional bank, you are asked to write a survey that asks clients about the levels of service they have received from the members of your bank's mortgage loan department.

- ***To report***
  As a writer for a national publishing company, you are asked to help write the quarterly report that presents your company's earnings and profit margins to its stockholders.

Identifying the purpose of a document may be a complicated task since many times a document has more than one purpose. Understanding that a single document might have multiple purposes helps you to write a better document.

For example, you may be asked to inform your organization's customers about a new life insurance policy. However, you know that your company also wants the customers to purchase the new policy. Therefore, when you begin writing the brochure or letter, you not only include the necessary facts that *inform* customers about the new plan, but you also include details that *persuade* these customers to purchase the new policy.

Connie considers purpose...

Connie understands that the new product brochure she will write must both inform her audience about the new product's functions and uses, and it must also attempt to persuade her audience to purchase the product.

Identifying the purpose or purposes of a document also helps you to understand more about its communication context. Audience is the final element we discuss to help you fully understand the communication context.

Response Exercise 5.2

Technical writers may write documents that have more than one purpose. Examine the example that follows and then read the list of documents below it. Imitate the example by inventing a possible writing *situation* for each listed document, and then writing at least two possible *purposes* for each document.

Example: A quarterly newsletter from a credit union mailed to its members (a) informs these members about new, current credit union policies and issues that affect them and (b) persuades them to attend the quarterly credit union meeting to elect new officers and discuss new policies.

1. Video cassette recorder owner's manual
2. Letter from a local charity organization
3. Organization's (business, government office, church) newsletter
4. Brochure for a local tourist attraction

Who's my audience?

While a document's situation and purpose may be relatively easy to identify, successfully pinning down its audience may be the most difficult task of all. However, understanding audience can make writing the document much easier. It helps to ask yourself a number of questions about audience before you begin to write the document:

Questions writers ask themselves about *audience*

- What is the type of audience: clients, customers, co-workers, supervisors?
- What previous knowledge does my audience have about this topic?
- What work do they do?
- What are their needs and interests in reading this document?
- What does my audience value? What are their biases?
- When, where, and how will my audience use this document?

Asking yourself these questions about audience before you begin to write is helpful for a number of reasons. The answers to these questions will help you to get a picture of that document's audience in your mind. Then you can more easily decide what information those readers may need—and what information they do not need—to understand your topic. This will help you decide exactly what information the document will include.

Next, the more you know about your audience, the more easily you can tap into and connect with that audience. You can determine how motivated your audience is and then shape your document to your readers' interests, or you can emphasize certain points in your document that tap into your readers' values or beliefs.

Last, the more you know about how your audience will use this document, the more competently you can write a document that those readers can use.

While answering these questions may not tell you everything you need to know about the audience, this information is still valuable. Establishing a general idea about the document's audience may lead you to ask more specific questions and to seek out the answers.

Finding out more specific information about your audience might be difficult, but taking the time to access it usually pays off. For example, you might locate demographic data about a specific audience type from your company's database, or you might use your experiences in writing to similar audiences to create a more complete picture of that audience.

Understanding the nature of the communication context is an important first step in the planning stage of writing documents. As a technical writer, you will find it nearly impossible to begin writing a document without knowledge of the situation, purpose, and audience of a given document.

Connie considers audience...

Connie has written other documents for Compu-Trends' customers before. However, this brochure is different because it is targeted not to *potential* customers but *current* customers—those who have purchased equipment from Compu-Trends in the past and are receiving the company's newsletter.

Connie accesses her organization's electronic database to find out more information about her audience. Identifying audience is Connie's last task before she begins to draft the product brochure.

Chapter 5 Task Exercise

Scenario

Imagine that you work as a technical writer for a small factory that produces and ships home audio equipment to retail outlets in your area. Your factory has devised a new, efficient method to package its equipment that reduces the likelihood of breakage during shipping.

You have been asked to write a manual for new employees who work in the shipping department. This manual will instruct these employees on how to complete this important, company-specific equipment process.

Task

1. Imagine that you are doing the preliminary planning for this manual. List the three elements that you need to understand about this document's communication context.
2. Using the concepts discussed in this chapter, describe in writing the process that you must go through to understand the document's communication context. Specifically, identify what questions you need to ask yourself and what you must know about the proposed document before you begin writing.

Chapter 6

Organizing and Outlining Documents

To write a successful document, organize your ideas before you begin to write. To organize, choose an organizing strategy that best suits your writing style and your writing situation, and then create an outline.

The second step in the planning stage of writing a technical document, after identifying the communication context, is deciding how to organize the information you will include. Organizing your ideas and information *before* you begin drafting will make the writing process proceed more efficiently.

To become skilled as a technical writer, you must be able to answer these three questions about organizing information:

- Why is it valuable to *organize* my information before I begin to write?
- What are the *strategies* I can use to organize my information?
- What do I need to know about creating an *outline,* and how can an outline help me organize my information?

After reading this chapter, you will be able to answer these questions regarding the value of organizing information, common organizing strategies, and the outline and its uses.

Why is it valuable to organize before drafting?

Once you understand as much as possible about the communication context of the document you are assigned to write, you must continue the planning process by considering organization.

As a technical writer, you will value the organization strategies introduced in this chapter. Many experienced writers often customize these strategies to suit their own writing styles and to accommodate different types of documents. To be a successful technical writer, you need to understand why it is important to spend time organizing your document.

There are three reasons why organizing information is an important part of the document planning process. Organizing your document helps you to generate ideas, assess your information needs, and map the document.

Brian's story...

Brian Thornton is a technical writer at a nonprofit organization, HelpCo, that provides free job search training to qualified men and women. Brian has been asked by Molly, HelpCo's manager, to write a letter to area businesspeople informing them of the upcoming "interviewing skills" seminar and to request their participation as "mock interviewers."

Not only must Brian inform his audience about the seminar, but he must also persuade them to participate. HelpCo will channel most of its available funds into conference materials, so the interviewers will receive little, if any, compensation.

Generate ideas

Organizing your ideas and the available information about the document may help you to generate more ideas about the document's subject. As a writer, you will discover that organizing your thoughts on paper makes you think about your topic in greater depth. Therefore, you are more likely to come up with new ideas and information about the document's topic. Generating ideas before you begin to draft the document makes the writing process more efficient and the document more useful to the audience.

Assess information

Organizing a document before you write allows you to assess the information needs of that document—that is, what additional information you need in order to begin writing the document.

In Chapter 5, we discussed how understanding a document's communication context helps you to decide what information should be included; moving through the organization stage is another way to discover what information you need.

Map the document

Organizing your document before you begin to write gives you a map that will help guide you through the writing process. While writing the document, sometimes you will choose to follow this map and other times you will choose to diverge from it. Either way, organizing your information before you begin to write gives you a structure that you can use *as you write.*

Brian knows the value of organizing...

As an experienced technical writer, Brian understands how valuable organizing is in the document planning process.

When asked about the importance of organizing strategies, Brian responded, "I spend time organizing each document I write. If I shortchange this time and begin writing too quickly my documents suffer, and I usually have to go back and organize anyway."

Organizing your document before you begin to write helps you to further plan the document by generating new, important ideas and by forcing you to discover what other information you need. Also, organizing information helps you during the writing process as well by guiding you through the main points of the document and allowing you to change and emphasize certain points along the way.

What organizing strategies can I use?

An organizing strategy is the method for choosing the way you sequence the information in your document. Technical writers must purposefully choose an organizing strategy for each document they write. While a particular organizing strategy may work well in one type of document, that same strategy may fail in another type of document. Therefore, there are several different types of organizing strategies that you can select.

To choose a strategy that is right for your document, you must ask yourself these questions about the document and about organization:

Questions writers ask themselves about *organizing strategies*

- What is my document's communication context?
- What strategy best suits my document and its context?

When technical writers are assigned a document to write, they learn about that document's communication context, they choose an organizing strategy, and they create a document outline. In this section you will learn about the six types of organizing strategies that you can choose from and the types of documents and communication contexts suitable for each strategy.

- Cause and effect
- Chronological
- Compare/contrast
- Divide and describe
- Order of importance
- Spatial

Cause and effect

The cause-and-effect strategy of organizing information enables you to describe both the cause and effect(s) of an event, idea, or recommendation.

Example: As a technical writer for an insurance agency, you are asked to compose a short letter informing your fellow employees about the effects of your company's recent company-wide policy banning smoking at the workplace. You begin the report by describing the *cause* (the non-smoking policy), and then you describe the *effects* of that event (increased productivity, cleaner work environment, more satisfied workers).

In any document using a cause-and-effect strategy, you may begin with the cause and then describe its effects, or you may begin with a series of effects and end with the cause. Writers need to choose the strategy most suitable to the document's communication context.

> **Brian chooses an organizing strategy…**
>
> Brian decides to organize his letter using a cause-and-effect strategy. He plans to begin his letter by describing the *effects* of the interview seminar, such as improvement in interview skills, more confidence, and better job placement success. Then, he plans to conclude the letter with a description of the *cause*, the job interview seminar. Brian believes that this organization strategy best suits this document and its communication context.

Chronological

A chronological organizing strategy allows you to describe information, such as a series of events or a set of tasks, as they occur or should occur in time. This chronological strategy is most often used when technical writers draft instruction manuals or write documents that describe a process.

Example: As a writer for a computer software firm, you are asked to collaborate on instructions for your firm's newest software program. You and the two software engineers who created the program use a chronological strategy to describe the use of the software in a logical, step-by-step manner.

Any document that must describe events or tasks in a sequential or chronological manner will benefit from this organizing strategy.

Compare/contrast

The compare/contrast strategy emphasizes the similarities and differences of the subject of a document, whether it is a particular topic, object, or event. Compare/contrast is an often-used strategy in a wide range of documents—from sales materials to correspondence and reports.

Example: As a writer for an automotive parts factory, you are asked to help write a report recommending the purchase of one piece of equipment for your factory's production line. Your report analyzes two pieces of equipment and recommends one that is the most cost-efficient to install and use. You decide that before you make the recommendation, the bulk of your report will describe the similarities and differences of the two pieces of equipment.

Many times, as in the example discussed above, you will use the compare/contrast strategy throughout an entire document. However, technical writers may also use this strategy within particular sections of a document.

Divide and describe

The divide-and-describe organizing strategy is useful for describing a process or an object. This strategy allows you to divide up the subject matter into its logical stages or parts.

When describing a process, you might divide it first into a series of major stages and then describe each of the smaller tasks that make up each stage. When describing an object, you might divide it first into its main parts and then describe the smaller parts that make up each main part.

Example: As a writer for an engineering firm, you are asked to collaborate on a description of a new cutting tool that your firm has designed. Eventually, this lengthy and detailed hardware description will help your firm to manufacture the tool. You divide description of the tool into its four major parts, and then you describe the smaller parts that make up each of these four.

The divide-and-describe strategy is useful for any document that must logically and efficiently describe a complicated process or a complex object.

Order of importance

The order of importance strategy organizes information by either increasing or decreasing order of importance. This, too, is a very popular organizing strategy, and it works well with a number of different types of documents.

In each document in which you use this strategy, you must decide whether to present the information in increasing or decreasing order of importance. In other words, you must decide whether to present the least important information first and conclude your document with the most important information or whether to present the most important information first and conclude your document with the least important information.

Example: As a technical writer for a bank, you are asked to help members of the mortgage loan department write a report that recommends a strategy for increasing their department's efficiency. You present the department's recommendation first (this is the most important information) and then describe the thinking that led up to and supports the recommendation (less important information).

Deciding on whether to use an increasing or a decreasing order of importance organizing strategy should be a decision based largely on your knowledge of the communication context. Most important in this case are the expectations of your audience. Ask yourself: Will my audience expect the most important information first or will they expect to be led up to this important information?

Spatial

The spatial strategy of organizing information is another option that you can use to describe an object. Using a spatial strategy allows you to describe that object as it exists in space: top to bottom, side to side, or inside to outside.

Example: As a technical writer for an advertising agency, you are asked to write a description of a new piece of equipment that workers within the design department will need to use. You choose to describe that equipment using a spatial strategy; you describe it logically from top to bottom.

A spatial organizing strategy is best if you are describing an object to users, but a divide-and-describe strategy may be best if you are describing an object about which others need to understand the shape or workings.

Understanding these six organizing strategies and the documents and communication contexts that are most suitable to each will help you to decide which strategy is best for each document you write.

After selecting an organizing strategy, most technical writers choose to create an outline of the document based on that strategy. In the next section, you will learn how an outline can help you in both the document planning and writing stages.

Response Exercise 6.1

As a technical writer you will have several different *organizing strategies* to choose from during the document planning process.

Read the list of documents below. Write down an organizing strategy that best suits each document and communication context. Then ask yourself why that strategy is most effective for the document and whether any other strategies would be effective as well.

1. Instruction manual for the users of the automobile
2. Description of a new cordless telephone for a design department that must design suitable packaging for the new product
3. Report describing the implementation of a new homeowner's insurance policy and discussing how the policy affects your company's overall policy sales
4. Report recommending a strategy for improving customer service based on a number of factors you and your team have identified
5. Description of the new office paper recycling process, recently instituted, for those personnel who must oversee the recycling process within their own departments
6. Description of two new accounting software programs and a recommendation of one that is best suited to your company's needs

How can an outline help organize information?

An outline is an ordered list of major points and supporting points that you believe will make up the document that you will write. As a technical writer, you create an outline for every new document that you produce, and typically you create an outline well before you begin to draft. You will find that it is difficult to begin a document, particularly a lengthy one, without using an outline as a starting point.

Commonly, technical writers use two different types of outlines: *sentence* outlines or *word* outlines. Both of these types are similar in that they illustrate the major and supporting points of a document, and in doing so both create a list, or hierarchy, of points.

The difference between the word and sentence outline is the level of detail that each includes. The word outline includes main and supporting points that are labeled with one word. The sentence outline, on the other hand, labels its points with lengthier phrases or sentences. When choosing either a word or sentence outline, be sure to use just one method throughout the entire outline. Being consistent in the type of outline you choose may make creating and following the outline easier.

As a writer, you may choose the type of outline that best suits your writing style; one type of outline is not necessarily more effective than the other. However, if you are writing a lengthier document, one that requires both headings and subheadings, you may find it useful to use a word outline to help you create the headings and subheadings you will use in the document. (See Chapter 9 for hints about formatting and placing headings and subheadings.)

While one type of outline is not necessarily more effective than the other, you need to understand what questions writers typically ask themselves about the quality and design of the outline they create.

Brian creates his outline...

Using a cause and effect strategy, Brian decides to create a sentence outline. He starts with the main points of his letter (beginning with the causes of the seminar) and attempts to include as much detail as possible while allowing himself enough flexibility to change points when he moves through the writing process.

Questions writers ask themselves about the *outline*

- Does my outline list the document's major points in the correct order?
- Does my outline include supporting points as well? Also, are these points in the correct order?
- Are the items I have listed either all words or all phrases?
- Is my outline detailed enough to be useful but flexible enough to change as I go through the writing process?

Outlines are valuable organizing tools that allow you to plan what points your document will include. Also, outlines provide you with a starting point, and you can use your outline to guide you through the writing process. Outlines may also help you to place and label the major sections of a lengthier document, particularly if you use the word outline as a guide.

Response Exercise 6.2

The outline is an important tool for planning and writing a technical document. Using what you have learned about creating outlines, write down the two ways that both word and sentence outlines help writers.

Next, think about a time when you used an outline to create a document, whether that document was a letter to an organization, an essay for school, or a family newsletter to send to friends and relatives. Now, respond to the following questions about that outline and its usefulness.

1. How much time did you spend creating the outline? Thinking back, was this time well spent? In what ways?
2. Did you refer to your outline as you wrote the document? If so, did you closely follow the outline, or did you change or revise it frequently?
3. If you have to write a document similar to this one in the future, will you use an outline again? If so, how will this outline differ from those you have written in the past?

As a technical writer, you will find that the document planning process is not complete without first organizing and outlining the information that you will include in a given document.

After understanding as much as possible about a document's communication context, you will now be able to select an organizing strategy that will best suit the document you will write. The writing process will be more productive and efficient if you first organize and outline your ideas and information carefully.

Chapter 6 Task Exercise

Scenario

Imagine that you work as a technical writer at the home office of a retail store that sells import items such as art, furniture, and clothing. The home office needs to contract with a new trucking company to deliver its items from the warehouse to its stores.

You have been asked to collaborate with a team of fellow employees to research three trucking company candidates, recommend one, and write a report about your findings. This report must describe the three trucking companies that your team has researched and clearly recommend one of them.

Task

1. You understand the communication context for this report, but you need to continue with the preliminary planning. List the steps that you need to move through in order to begin writing the report.
2. List the organizing strategies discussed in this chapter, and write down the strategy you believe is most suitable for this report. Why do you believe that strategy will be most effective?
3. To begin writing the report, you and your team need to prepare an outline. Using what little information you do have, create a possible outline for this report.

Chapter 7

Locating and Using Common Information Resources

To become a good technical writer, you need to understand how to locate more information about your topic. In this chapter, you will learn about the most common information resources and ways that writers use them.

As a technical writer, you often spend as much time locating information about the subject of a document as you do organizing and writing that document. The kinds of information that you will need to access are as different as the documents you write. However, several information resources are quite common to most organizations, and these resources help you to locate the information you need.

In this chapter, you will learn how technical writers locate the information they need in order to begin writing a document. At the end of this chapter, you should be able to answer the following questions about locating information.

- What are the most common *information resources* that I will use?
- What are the best ways to *use* these information resources?

While you may not need to be familiar with all of the information resources we discuss in this chapter, and there may be different types of resources available at your workplace, the details presented here provide a solid starting point for understanding how to locate information.

What are the most common information resources?

Once you understand a document's communication context, choose the most suitable organizing strategy, and begin an outline of that document, you may realize that you do not have all the information you need to begin writing. Before you can proceed, then, you must find out more information about your document's topic.

As a technical writer, you have several different information resources at your disposal. If you know where these resources are located and how to use them, you can find the information you need and begin the writing process. In the next section, we describe six of the most typical information resources that you will use as a technical writer:

- Your experience
- Interviews
- Request and response letters
- Surveys
- Objects or processes
- Internal resources

What are the best ways to use these information resources?

This section describes six information resources and offers strategies and tips for accessing information from them. Understanding the purpose of each resource and how to use it is important for any technical writer.

LaRhonda's story...

LaRhonda Pearson is a technical writer at a large cosmetics company. LaRhonda has been asked by Geraldine, her supervisor, to collaborate with two of the company's dermatologists on a brochure to accompany a new line of skin care products.

Geraldine tells LaRhonda that the brochure will provide skin care representatives with in-depth product information. The brochure will need to communicate complicated, scientific information in a clear style that any skin care representative with a limited science background can understand.

Your experience

Your personal experience is an important information resource, and it is a useful place to begin, since it's your most readily available resource. Consider that this information resource contains both your *past* and *current experience.*

Remembering and drawing upon your *past experience* is a useful place to start if you need more information about a topic. Try to think of any experiences you have had writing a similar type of document or working on a similar project. Past experiences such as these can be valuable sources of information.

For example, to refresh their memories about past documents or writing projects, technical writers often keep archives, or files of completed documents, close at hand. These archives provide them with useful information about each document they have prepared, including its subject, audience, and purpose.

Your *current experiences* relevant to the document's topic may be valuable to use when writing the document. For example, within your organization you may observe current events that may be useful to include in your document. Before including these details, you may want to corroborate your ideas or experiences with the experiences of your co-workers.

LaRhonda considers her experience...

LaRhonda has collaborated on projects like this one in the past. She remembers a similar type of information brochure she wrote for a line of lip-care products.

LaRhonda looks through her archive of past documents and finds the lip-care brochure she wrote last year. Also, she discovers a set of meeting minutes that she wrote during one of her team's document planning meetings.

As you can see, your past and current experiences can be valuable information resources. To best access this kind of information, ask yourself these questions:

Questions writers ask themselves about *experience*

- What past experience do I have with the topic of this document? How can I access this information?
- Have I written this type of document before? If so, how can I access this information?
- What current experiences am I having relevant to this topic?
- Are any of my colleagues having similar or different experiences?
- How can I access information about these current experiences?

Interviews

Another information resource that is useful and relatively accessible is the experience and knowledge of *other* people: co-workers, customers, clients, or anyone who has the information that you need. One of the most effective and productive ways to access information from an individual is through an interview.

An interview, whether it is conducted face to face or via telephone, allows you to access the information you need directly from its source: the individual. An effective interview, one that gives you substantial information, must pass through three stages: preparation, interview, and follow-up.

Interview preparation

To prepare for the interview, *identify your information source.* Once you have identified the persons you need to interview, you must contact them by telephone, e-mail, or in person to discuss the possibility of scheduling an interview.

But before you schedule the interview, prepare the *interview purpose statement.* Write down the (a) purpose of the interview, (b) use you will make of the responses you receive, and (c) approximate amount of time you will need for the interview. When you schedule your interview, make sure you communicate these three points to your interview subject.

Once you have scheduled the interview, you need to prepare your *interview script,* or the series of questions you will ask. Always prepare a script for the interview. Successful technical writers ask themselves these questions about the interview script:

Questions writers ask themselves about an *interview script*

- Does each response require more than a simple yes or no answer?
- If a yes/no question is asked, does it follow up with a question requiring more discussion?
- Does each question encourage a focused response?
- Does each question ask about one topic at a time?

Interview

Be sure to arrive on time and prepared at the selected interview location. One important way to be prepared is to remember to bring along these three items:

- A *watch* to make sure you keep the interview within the predetermined time
- A *notepad* and at least two pens to take notes
- A *tape recorder* to tape the interview (ASK PERMISSION FIRST)

When you arrive for the interview, make sure to introduce yourself and summarize the interview purpose statement. Remember that you have primary control over the pace of the interview.

Tips writers keep in mind about the *interview*

- Give the individual enough time to respond to each question—do not interrupt.
- If the individual's response is not clear or too vague, ask for an explanation or an example.

- If the individual rambles, keep the response focused by restating the question.

Conclude the interview by thanking the individual, and be sure to arrange for a follow-up interview, if necessary.

Interview follow-up

After you have conducted the interview, send the individual a thank-you note. Also, if a follow-up interview is necessary, either in person or via telephone, be sure to conduct the interview in the same courteous manner that you did the initial interview.

LaRhonda conducts an interview…

LaRhonda discovers that in order to begin writing her brochure she must understand how a skin care representative consults with a new client.

LaRhonda schedules an interview with an experienced skin care representative who will eventually sell the new product line. After a thirty-minute interview, LaRhonda has enough information about a typical consultation to help her decide how to organize the details about the new skin care products in her brochure.

Request and response letters

Another information resource that helps you to access the experience and knowledge of other people are letters of request and response. Like an interview, request and response letters—conventional letters and e-mail messages—are very effective ways to access information from an individual. Unlike an interview, however, this type of correspondence allows you to get valuable information without actually meeting with the individual. This could be advantageous if you are dealing with time constraints or distance issues.

Letters of request and response are an efficient and inexpensive way to access information. By storing the letters that you send and receive, you will have an instant archive of information.

Listed below are a number of strategies that allow you to receive information successfully through letters of request and response.

Tips writers keep in mind about *request and response letters*

- Begin the request letter by identifying why the recipient is an important information source and how you will use the responses.
- Ask questions that are open ended, but focused enough to generate productive responses.
- Conclude by courteously stating a response deadline.
- Once you receive the response letter, read it through carefully and note any confusing or vague responses.
- If necessary, write a second request letter that asks the recipient to clarify or further explain points that are unclear.

Response Exercise 7.1

Knowing how to write a *letter of request* is an important and useful skill. Choose one of the scenarios below and write an outline for a letter of inquiry. In your outline, include a specific list of questions. Use what you have learned about letters of request to create your outline.

Scenarios *(in each case, you choose the vehicle, event, or service)*

1. As a prospective vehicle buyer, you need more information about a particular vehicle's cost, performance, and efficiency.
2. As a possible attendee, you need more information about the schedule, events, and purposes for a summer festival in your area.
3. As a likely customer, you need more information about the cost, schedule, and services provided by a home or lawn care service of a company in your area.

Surveys

One of the best ways to access information from a large group of people is by sending them a survey to complete. Surveys can be one page or multiple pages of questions that ask individuals for their input or responses regarding an event, issue, or product.

Typically, surveys are useful for identifying the reactions and responses of a large group of people, since a one- or two-page survey can be sent to hundreds of people at a relatively low cost. However, they also may be sent to small groups of people. Whether there are five or five hundred respondents, a well-designed survey is a useful and inexpensive method for accessing information.

As a technical writer, you may choose to write a survey to find out more information about a topic, product, or event so that you may begin writing a brochure, report, or other document. This will require asking yourself several important questions about writing an effective survey.

Questions writers ask themselves about *surveys*

- Do I have a target audience, or set of respondents, in mind for this survey?
- Have I asked an effective set of questions that are focused enough to generate specific responses yet open ended enough to allow for descriptive responses?
- Have I administered my survey to a test group of respondents before I send out the survey to the actual respondents?

Objects or processes

One important way to understand as much as possible about the subject of a document is to locate and examine that topic yourself; this method is particularly useful if the subject of your document is an object or a process.

Technical writers frequently create documents that describe what objects look like or do or that describe how processes work or run. Hardware descriptions, which are the most common of these documents, are usually written after technical writers have examined the equipment, tool, or other object. These descriptions are often included in user's manuals or instruction booklets.

Process descriptions are the most common documents written about processes. (Process descriptions may also be included in *instructions*; see Chapter 3 for more information about instructions.) Technical writers observe and describe a process and include a description of that process in reports or other documents.

To begin either a hardware or a process description, you need to locate and examine that object or process. You can also use this technique for documents other than hardware or process descriptions.

For example, a report recommending the purchase of one piece of equipment might contain a section describing that equipment. In this case, to best convince your audience about the value of that equipment, describe that equipment thoroughly. Listed below are a few of the most important tips to keep in mind when you access information by examining an object or process.

LaRhonda examines an object...

To best convey the product line information to her audience, LaRhonda decides to examine each of the products in the new skin care line.

During one of their team meetings, Janine, one of the two dermatologists collaborating on the brochure, gives LaRhonda a demonstration of each of the products.

Tips writers keep in mind about *examining objects or processes*

- Obtain permission to examine, observe, photograph, sketch, or videotape the object or process.
- Ask questions about the object or process. If possible, direct your questions to those people who know the most about the object or process.
- Ask to participate in the process or to handle the object, if possible.
- Take detailed notes, photographs, and/or sketches during your observation or videotape the object or process and the questions you ask.
- Label your notes, sketches, or other materials so that you may refer to them quickly and easily.
- Retain your notes and other materials about the object or process even after the writing project is finished.

Response Exercise 7.2

Understanding how to *examine* an *object* or a process is an important task in locating information. To practice observing an object, choose an item, one that you have readily available, from the list below:

- Hammer
- Flashlight
- Screwdriver
- Clock (battery operated)
- Hairbrush
- Egg beater (hand-held)

Find the object and write down your responses to the following prompts. These responses will help you to describe the object. You may use these prompts to help you describe nearly any object.

1. Name and identify the two or more main parts of the object.
2. Examine each main part. Do these main parts contain smaller parts? If so, name and identify these smaller parts.
3. Examine and describe each main part and, if necessary, its smaller parts. As you describe each part, be sure to include these details:
   - Function of the part
   - Color, texture, and/or type of material of part
   - Dimensions (length, height, and width) of part

Internal resources

Internal resources are information sites found within your organization. These resources may be a *paper archive* (several file cabinets or a whole room) or an *electronic database.* Internal resources, whether stored in an archive or electronically, provide you with valuable, company specific information.

The information typically found in an internal resource site may only be available from your organization. This information may be about members of your organization, the products or services your organization provides, or economic or financial information about your company.

Internal resource sites may include facts, data, and reports regarding the following topics:

- Personnel information
- Financial information
- Company services information
- Company products information

Your company's internal resource sites may also contain information about the products or services that are provided by other companies. For example, your company might have information about the services or products of a company that has worked with your company before. Internal resources, whether they are stored in a paper archive or electronically, are valuable sites to find information about nearly any topic regarding your company.

To be a successful technical writer, you must be able to identify, locate, and use the six information resources we discussed in this chapter. While you may not use every resource for each document that you write, understanding which resource is most useful for which type of project is valuable information.

LaRhonda uses an *internal resource...*

To prepare for a meeting she will have with her team, LaRhonda accesses one of her company's most useful internal resources: the electronic database.

Using her computer, which is linked to the company's database, LaRhonda finds information about the manufacture of the skin care products and the method of product shipment. This information, along with all the other information she has accessed, allows her to understand a great deal about the new product line.

Chapter 7 Task Exercise

Scenario

Imagine that you are a technical writer for a large, multi-national chemical corporation that ships its products all over the world. As an experienced member of your corporation, you will help conduct a training seminar for new technical writers. You have been asked to lead a discussion about the purposes and uses of your corporation's information resources.

Assume that the new technical writers must become familiar with all six of the information resources presented in this chapter. Respond to the questions below to begin an outline of your portion of the training seminar.

Task

1. List the six information resources that will help writers locate more information.
2. After each resource that you have listed, write down your responses to the following questions.
   a. Explain the purpose of this resource. What is its most valuable use or function?
   b. When should this resource be used? Describe a task or project that might consider this information resource valuable.
   c. Identify at least two strategies or tips that the writer should understand about using this resource. Why is understanding these strategies important?

Chapter 8

Incorporating Visual Aids Into Documents

To become a good technical writer, you need to know when information needs to be communicated visually. This chapter will discuss why visual communication is important and how to create visual aids and place them into documents.

Until now our discussion of the document planning process has focused almost exclusively on ways to generate, organize, and outline written information. However, one important part of the document planning process includes understanding visual communication and ways to use visual aids within your documents.

Visual aids are elements in a document, such as charts, graphs, tables, and illustrations, that communicate information visually. When a range of information needs to be conveyed in a document, visual aids can help to clearly present that information.

In this chapter, we will cover why visual aids are important and how to create them and effectively place them within a document. As a successful technical writer, you will need to know how to respond to these questions about visual communication:

- When do writers *use* visual aids?
- What *types* of visual aids do writers use?
- What *strategies* do writers adopt to use visual aids effectively?

When do writers use visual aids?

Often during the document planning process or even during drafting itself, technical writers identify places within the document where visual aids will help to communicate points or ideas. Pinpointing the places in a document where visual aids will be most useful is an important skill for a technical writer to have.

As a technical writer, you need to understand and remember four points about visual aids and their placement within a document. Successful writers know the responses to these important questions about the use of visual aids.

Questions writers ask themselves about the *use of visual aids*

- Will the information be *clearer* and more *understandable* if I use a visual aid?
- Will my inexperienced audience, who has never been exposed to this information before, *learn* more from a visual aid?
- Will the *interpretation* of the information be more *accurate* if I use a visual aid?
- Will the information I present in a visual aid *complement*, not stand in for, the discussion I am having?

First, technical writers strive to make their written text as *clear* and *understandable* as possible, and many times a visual representation of an object, process, or idea is a more effective way of communicating it.

For example, Figure 8.1 accompanies Step 5, "chemical paste application," in an experiment instruction guide for middle school science students. Providing this visual aid is a useful and effective way to clarify this step in the experiment. This visual aid is useful because it illustrates a process—chemical paste application—that these students have never performed before.

Figure 8.1 A step in a process is illustrated
Chemical Control of Plant Growth. USDA. Science Study Aid #7. USGPO. 1972, 12.

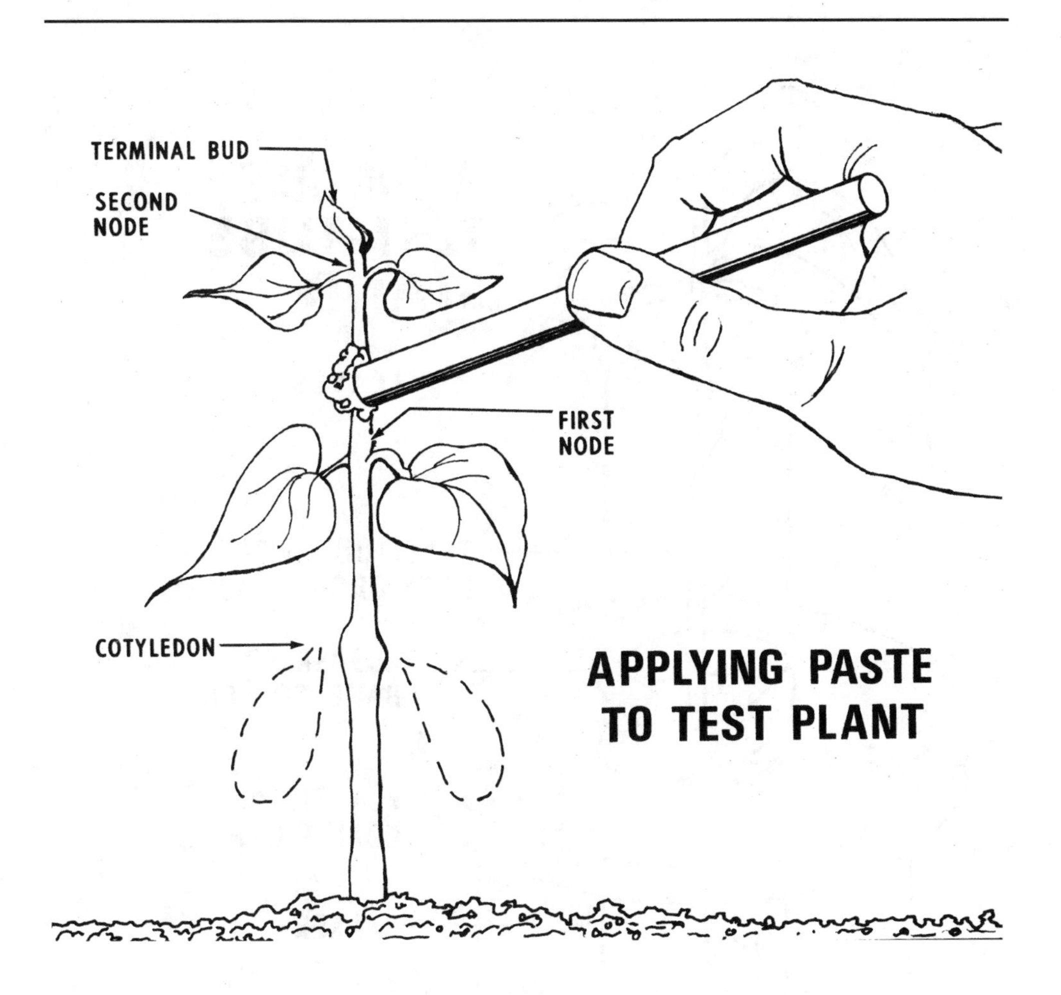

Second, using visual aids within a document may help those readers who are inexperienced with a concept or idea understand or *learn* about it more easily. Readers who do not know a great deal about the subject being discussed may learn more readily if a written discussion is accompanied by a visual aid.

For example, the visual aid in Figure 8.2 is carefully labeled and clearly identifies the different parts of a test-tube stand. Labeling visual aids is critical because readers need to understand the important components of any illustration. The middle school science students who refer to this figure and its accompanying text will be using a valuable learning tool.

Figure 8.2 A clearly labeled visual aid

Chemical Control of Plant Growth. USDA. Science Study Aid #7. USGPO. 1972, 16.

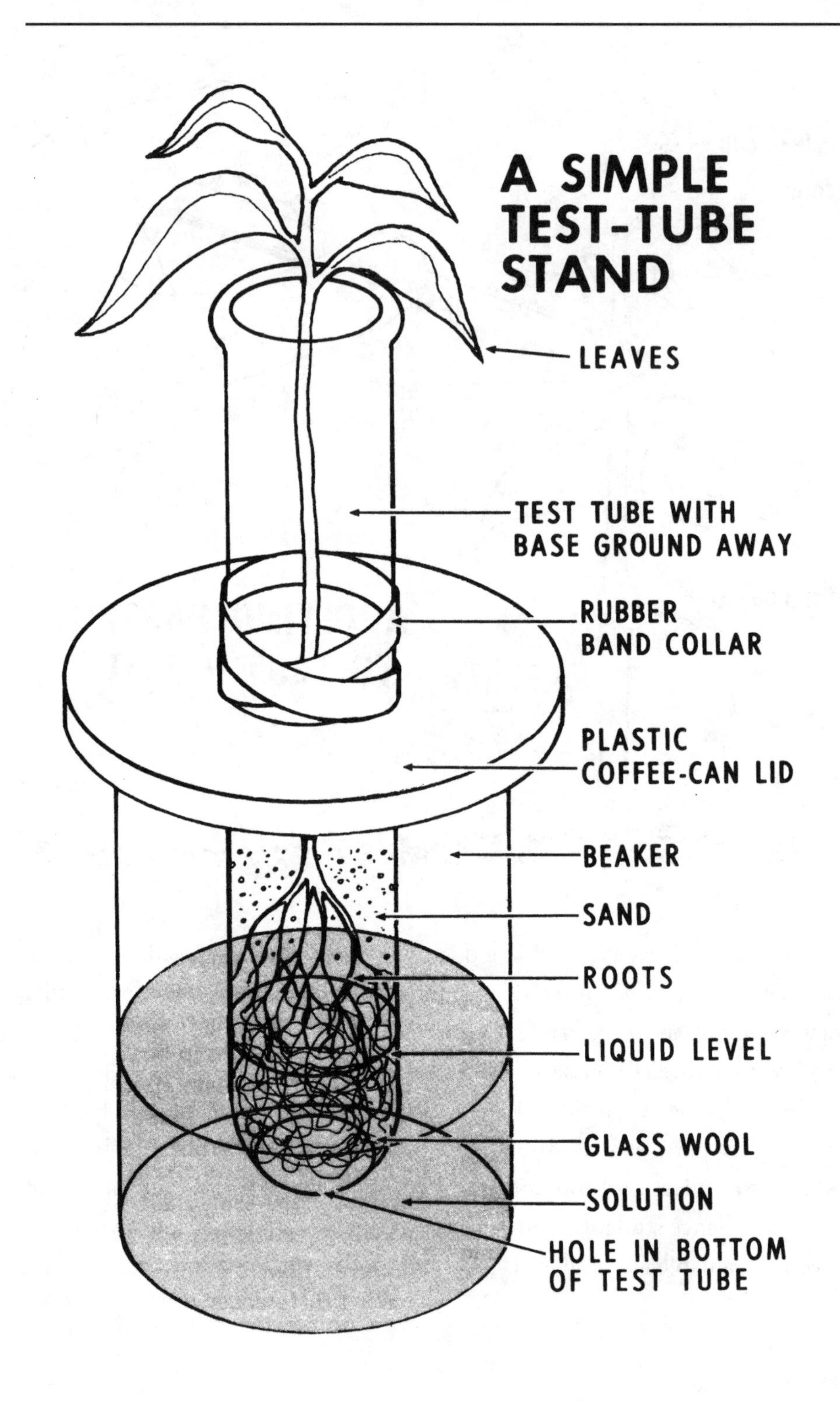

Third, visual aids can help readers understand and *interpret* information in a more *accurate* fashion. For example, the information presented in Figure 8.3a, the results of a soil survey, is nearly impossible to scan quickly since its paragraph organization does not help the reader accurately identify, compare, or understand the information.

Figure 8.3a Paragraph describing soil survey results

The loam soil texture has a pH of 5.6%, organic matter capacity is 3.7 megs, and cation exchange is 14.4. The loamy sand soil texture has a pH of 5.65, organic matter capacity is 4.03, and cation exchange is 12.1. The sandy clay loam soil texture has a pH of 6.1%, organic matter capacity is 4.03 megs, and cation exchange is 12.1. The sandy loams soil texture has a pH of 5.63%, organic matter capacity is 4.08 megs, and cation exchange is 11.4.

In contrast, Figure 8.3b presents the information in a visually accessible way, allowing the reader to understand, compare, and interpret the data much more quickly. Even though both the paragraph and the table present the same information, interpreting the information presented in the table is much easier.

Figure 8.3b Table showing soil survey results
Nursery Pest Management: Draft Environmental Impact Statement. USDA Forest Service. Oct. 1991, III–10.

	Average Values or Conditions		
Soil Texture	***pH (%)***	***Organic Matter Capacity (meg)***	***Cation Exchange***
Loam	5.6	3.7	14.4
Loamy sand	5.65	4.03	12.1
Sandy clay loam	6.1	4.03	12.1
Sandy loams	5.63	4.08	11.4

Fourth, visual aids are important parts of a document that may make reading and understanding the information that is being presented easier. However, you must understand that visual aids are almost always a *complement* to the written text and not a *substitute* for it.

John's story…

John Miller is a technical writer for a large construction company, Stein Industries, that builds commercial properties such as hotels, restaurants, and office complexes. John's supervisor tells him that he will work collaboratively with project managers and a supervisor on a high-priority project. John's team will create a proposal that will estimate and bid on project costs, schedule, and personnel.

John has collaborated on projects like this in the past, and he leaves the first team meeting with a useful outline, a set of data, and the task of fleshing out the outline for the team.

Therefore, the best visual aid is one that works with the surrounding written information to help your audience better understand the subject. You need to spend time creating visual aids and placing them within documents; however, do not shortchange the discussion that you have about those visual aids, since visual and written information should work together.

Response Exercise 8.1

Knowing when to use visual aids in a document is an important skill that successful technical writers understand. To better determine when visual aids are necessary within a document, write down your responses to the following questions.

1. During the document planning process, what four questions regarding the use of visual aids must writers ask themselves?
2. For each strategy, describe in writing why it is important to understand this strategy when deciding about visual aid use.

What types of visual aids do writers use?

Once you have determined that a visual aid is necessary in a document and that it will help the audience understand a point or concept, you must choose the type of visual aid that will be most useful. Writers have several different types of visual aids to choose from; they choose the appropriate visual aid by asking themselves these questions:

Questions writers ask themselves about *types of visual aids*

- Why is a visual aid necessary here? Will a visual aid make this concept more clear, make my audience learn more easily, or make the information interpretation more accurate?
- What function do I want the visual aid to perform?
- What visual aid best suits the function that is necessary here?

To respond to these questions, you must first understand the different types of visual aids and the kinds of functions that they perform. Successful writers know how to choose the right visual aid to suit the appropriate purpose, and they usually select from six main types of visual aids, each of which performs a different function:

- Tables
- Graphs
- Organizational charts
- Drawings
- Photographs
- Maps

John plans what visual aids to use...

Taking into consideration the communication context as well as the information being presented, John decides that his team proposal will need several different types of visual aids to best communicate the information.

John uses the outline, his meeting notes, and his prior knowledge of proposal writing to begin choosing the visual aids he will use.

In this section we discuss each of these visual aids by describing the most common purposes for each and the functions that they typically perform in a technical document.

Tables

Tables are visual aids that organize data into columns and rows. Tables are most useful for displaying large amounts of numerical data. By examining Table 8.4, a table that illustrates a garden tool inventory from a hardware store, you can see that besides just organizing data, tables can be useful for *comparing* data as well.

Table 8.4 enables the reader to make a quick comparison of the number of in-stock items with the number of those on order. It also facilitates a totaling of the spring inventory of each item.

Table 8.4 In-stock and on-order hardware inventory

	Shovel	*Trowel*	*Hoe*
In Stock	17	28	15
On Order	3	5	5
Total Spring Inventory	20	33	20

Although the table is most useful for displaying numerical data, non-numerical data may be presented effectively in a table as well. For example, see Table 8.5, which clearly displays the non-numerical data of tool type, manufacturer, and season.

Table 8.5 Equipment stocking information from hardware inventory

Tool Type	*Manufacturer*	*Season*	*In Stock*
Hoe	Garden Smart	Spring	Yes
Leaf Rake	Garden Smart	Fall	Yes
Shovel	ToolWay	Spring/Fall	No

Graphs

While tables organize data into simple columns and rows, graphs can organize data in three ways: according to time (using a line graph), according to amount (using a bar graph), or according to percentage (using a pie graph).

Line graphs are most useful for showing trends over time. In the line graph in Figure 8.6, notice how the data indicates the frequency of help desk use over a period of thirteen weeks.

> **John uses graphs...**
>
> John decides to use two types of graphs within the document: a line graph and a bar graph.
>
> The line graph shows the projected revenues that the audience will gain from this project, while the bar graph compares the data from the audience's company and their competitors regarding profitability.

Figure 8.6 Line graph illustrating use over time
Denning, Tracey. "1989." *Customs Today*. 24.3. Dept. of Treasury, 27.

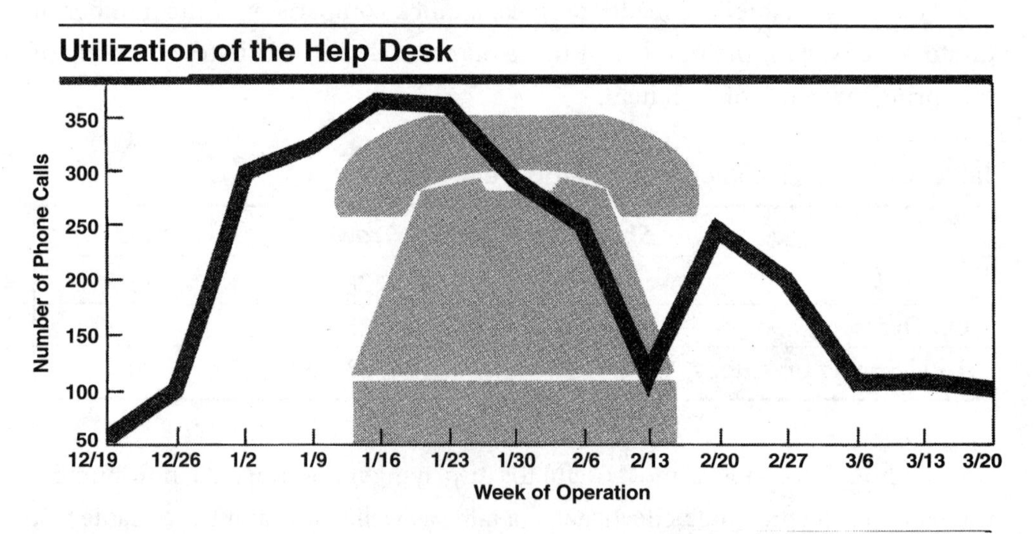

The bar graph organizes data by using columns or bars to display and compare different quantities. Bar graphs show amounts of the same item at different moments in time. Also, bar graphs, such as Figure 8.7, show quantities of different items at the same moment in time. Finally, bar graphs illustrate how different parts of an item make up the whole.

Figure 8.7 Bar graph illustrating breakdown of sites

"Updated Data Show Bioremediation Trends." *Bioremediation in the Field.* USEPA. EPA/54/N-95-500 #12. Aug. 1995, 5.

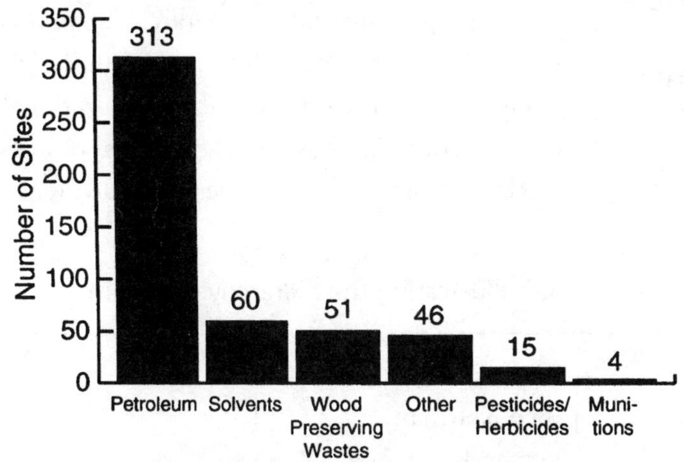

Breakdown of sites by type of contamination.

The last type of graph, the pie graph, best illustrates the various percentages that make up a whole. Figure 8.8 shows a variety of percentages that comprise the whole. This visual aid is useful for comparison.

Figure 8.8 Pie graph illustrating percentages of a whole

"Updated Data Show Bioremediation Trends." *Bioremediation in the Field.* USEPA. EPA/54/N-95-500 #12. Aug. 1995, 6.

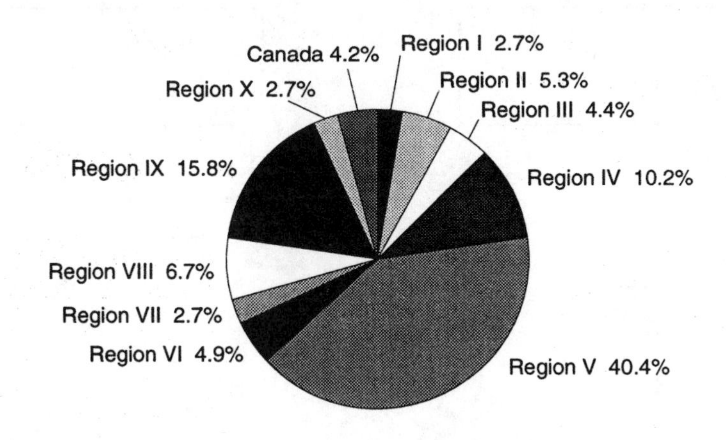

Distribution of Bioremediation Projects by Region.

Organizational charts

Organizational charts are useful visual aids that clearly represent the organization, or hierarchy, of a company, department, or project team. These types of charts normally use boxes and lines to depict organization, and the charts are useful for showing hierarchy and depicting the lines or channels of communication within an organization.

Figure 8.9 is an organizational chart that might be placed into a document, such as a proposal or a recommendation report, that must clearly indicate the chain of command and hierarchy in an organization to those readers who are unfamiliar with it. Notice how easy it is to scan the organizational chart and to understand how one department or individual is related to another department or individual.

> **John uses an organizational chart...**
>
> In the section that describes John's company, and specifically the people who will be involved in this project, John chooses to use an organizational chart.
>
> This chart will show the hierarchy of the project members and how the lines of communication flow from the team to the company.

Figure 8.9 Organizational chart illustrating the hierarchy of a team

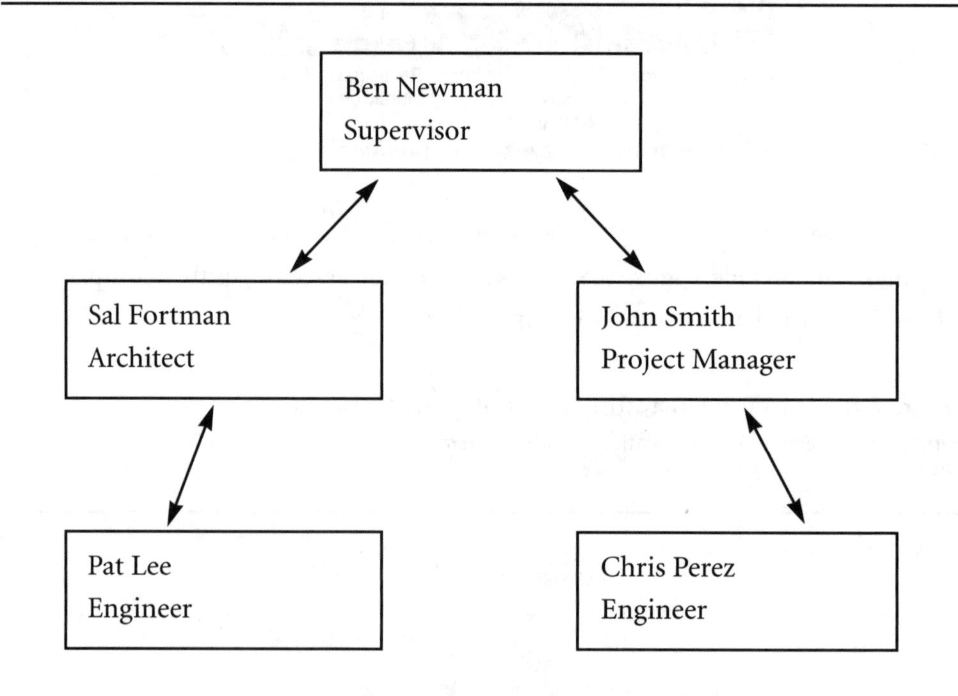

Drawings

Drawings realistically portray objects and are most useful for showing detail or highlighting a certain point of view. Drawings are visual aids that may illustrate a variety of objects, such as a roadside exhibit stand (Figure 8.10), an elevation of a structure (Figure 8.11), or the cut-away view of a mountain range (Figure 8.12).

Figure 8.10 A drawing of a roadside exhibit stand
"Portable Exhibit Bases." *Wayside Exhibits Users Guide.* NPS 1997, 17.

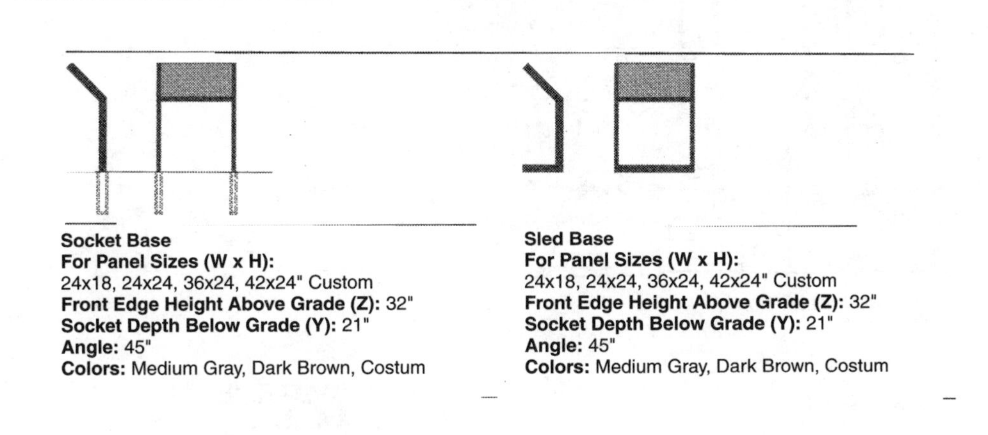

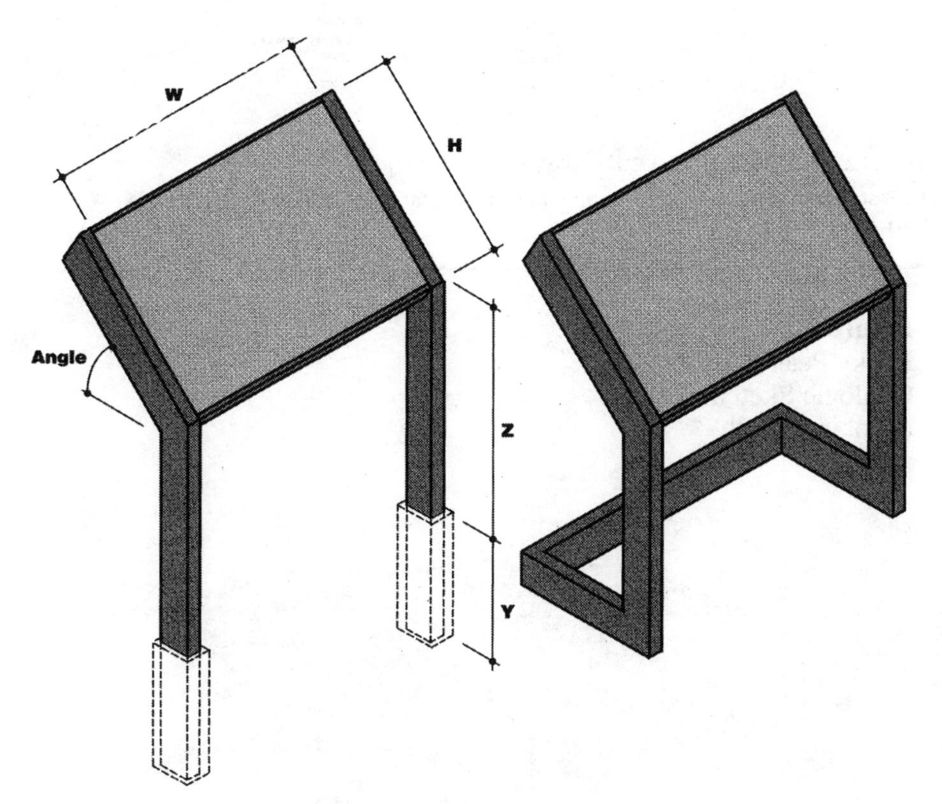

Figure 8.11 A drawing of an elevation of a home

"Enston Homes." *HABS/HAER Review*. FY1993 NPS, 59.

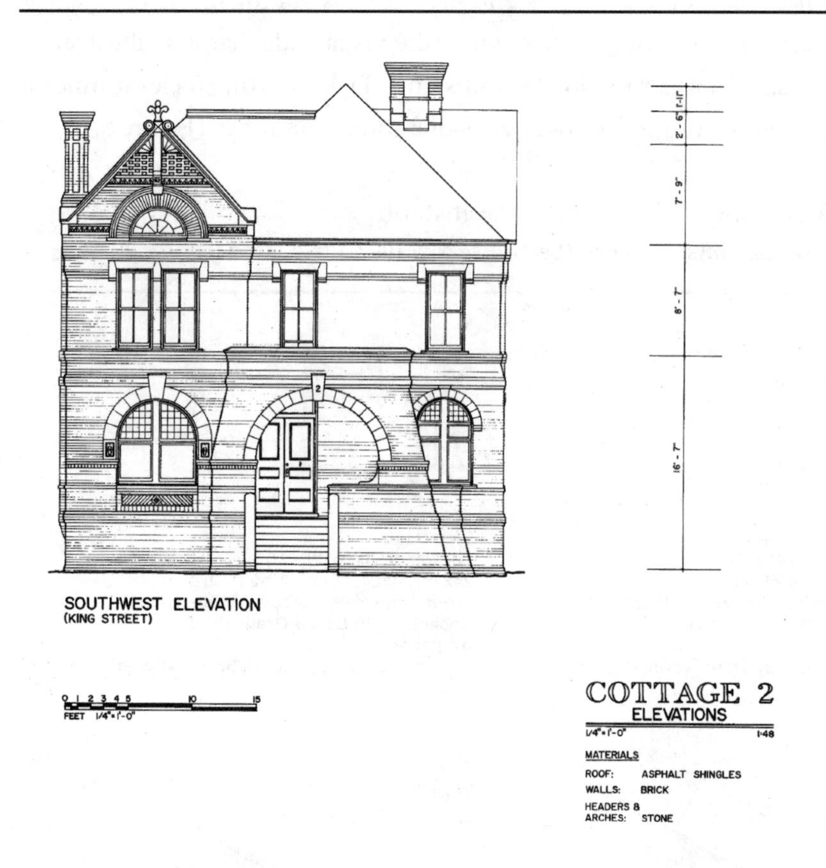

Figure 8.12 A drawing of a cut-away view of a mountain range

Nursery Pest Management: Draft Environmental Impact Statement. USDA Forest Service. Oct. 1991, III–6.

Figure III-2
Lucky Peak Nursery
Geologic Sketch
(Looking North)

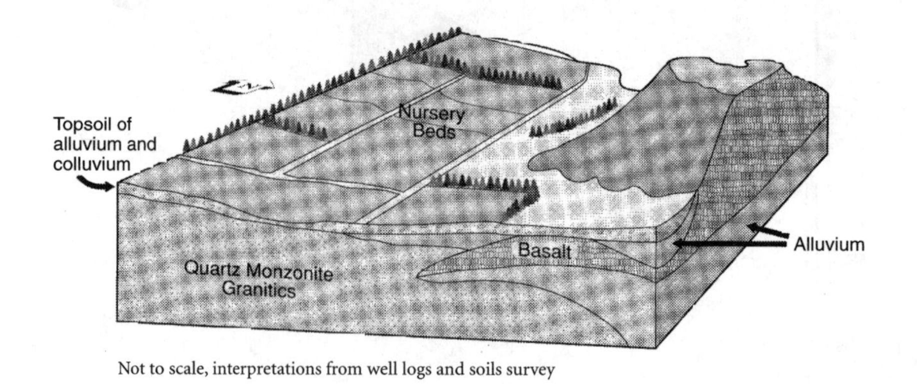

Not to scale, interpretations from well logs and soils survey

Notice how by presenting a certain point of view (the cut-away view), Figure 8.12 emphasizes the grades of the mountain and the different soil types that it is made up of.

Photographs

Photographs, like drawings, can depict a wide range of objects; however, photographs are, of course, more realistic than drawings. Photographs, like the one in Figure 8.13, are most useful when the reader needs a more exact representation.

Figure 8.13 A photograph illustrating a system inspection pilot at work
Airspace System Inspection Pilot. GS-2181-9/11/12. Announcement #FAA-ASIP-01. Nov. 1989, 2.

Figure 8.13 depicts an airspace system inspection pilot running the controls at his workstation. Photographs, as you can imagine, may capture a number of different objects, events, or people.

Maps

Maps are the final type of visual aid that technical writers frequently use in their documents. They are effective for showing the geographical or "built" features of a region. Maps, like the one in Figure 8.14, can be placed into the text of the document itself or, if larger, can be separated from the body of the document and placed in the appendix.

> **John uses a photograph...**
>
> To best show the details of work that was previously performed at a construction site, John includes a photograph of the equipment and personnel in action.
>
> John will provide a context for this visual aid by describing how his company will work well for them.

Figure 8.14 Map of a Grand Teton nature trail
Hermitage Pt., Two Ocean and Emma Matilda Lakes Trails. NPS. USGPO. 1997-575-167, 2.

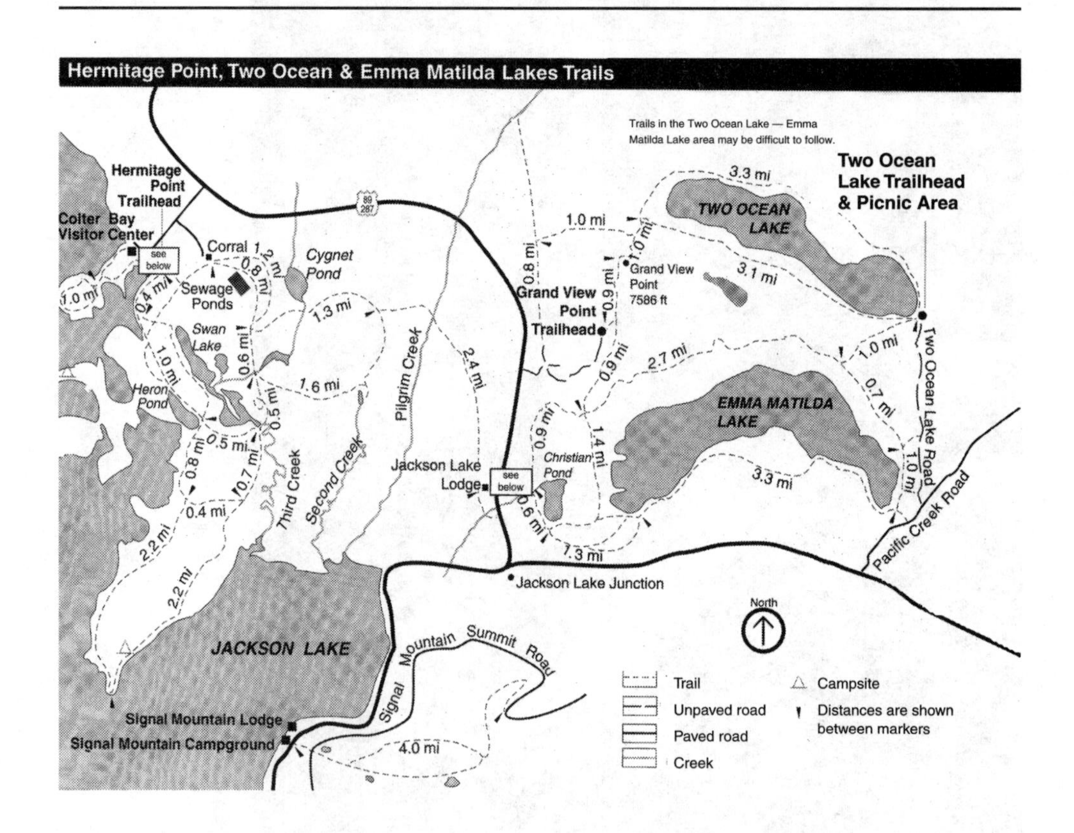

All six of these types of visual aids are useful for certain tasks. Understanding that each visual aid has a unique purpose makes selecting and using them within documents much easier and more effective.

Response Exercise 8.2

Understanding the function of each of the six main types of visual aids will help you to select and use these visual aids more effectively in the documents that you write.

Find a document—a brochure, user's manual, newsletter, letter, report, or another type—that uses any of the six types of visual aids discussed in this chapter. Answer the following questions about each type of visual aid. The more documents and visual aids you use for this exercise, the more you will understand about the variety of uses for visual aids.

1. What type of visual aid is used? Briefly describe the written text that accompanies the visual aid.
2. Indicate, in a sentence or two, the function of this visual aid. Does it compare numerical data? Realistically illustrate an object or process? Show a trend over time?
3. Describe in writing how your understanding of this section of the document would change if the visual aid were not there.

What strategies do writers adopt to use visual aids effectively?

Successful technical writers understand that certain strategies or conventions are used for each type of visual aid to make it more functional in the document. In this section, we discuss the strategies that you must understand about creating, placing, and using visual aids within the documents that you write.

- *Provide a context*
  Make sure the text clearly explains why the visual aid is needed. Basically, you need to discuss the visual aid by providing background or context for it before you place it into the document. The context that you provide for the visual aid depends on your document's communication context, and especially on your audience. Depending on your audience—their skill level, education, and prior experience with the subject—you must provide either more or less detail.
- *Present one type of information*
  Present only one type of information per visual aid. If there are two or more competing types of information, separate them into different visual aids.

- *Be consistent*
  Be consistent in the way that you use labels and terms. If you identify a part by one label in the visual aid, be sure to refer to it by the same name when you discuss it in the text.
- *Refer to visual aid*
  Number each visual aid in the order that it is presented in the document, and title each visual aid according to its subject or topic. Refer to each visual aid in the text by its number, either parenthetically or within the sentence.
- *Place visual aid effectively*
  Attempt to place the visual aid as closely as possible to the text that first discusses or provides the context for it. When you place the visual aid into the document, make sure that it corresponds to correct design procedures. (See Chapter 9 for more detailed discussion about document design and layout options, including tips on layout options for visual aids.)

John successfully uses visual aids...

Visual aids are important in John's proposal, and he has carefully selected the different types of visual aids he will use.

John reviews the strategies that are important to creating and placing visual aids before he begins to further revise his team's outline of the proposal.

Understanding the value and function of different types of visual aids, knowing where to place them in the document, and recognizing the best strategies for creating them are important skills for successful technical writers to use.

The descriptions and strategies that you have learned in this chapter are good starting points for understanding the importance and usefulness of visual communication in technical documents. The more documents needing visual aids that you write, the more accomplished you will become at placing, creating, and discussing those visual aids.

Chapter 8 Task Exercise

Scenario

Imagine that you work as a technical writer for a small firm that specializes in repairing electronic items such as computers, video cassette recorders, and televisions. Soon, your company will be investing in new word processing software, and this software must be capable of creating and designing visual aids in addition to having standard word processing capabilities for written text.

As your firm's technical writer, you will use the word processing software most frequently. Your manager, who will decide what software to purchase, needs your input. Specifically, you must respond to these questions about the visual aids you use on the job.

Task

1. Identify each of the six different types of visual aids you may create in your work at the firm. Describe the function of each type of visual aid. Does it show the percentages that make up a whole? Does it compare numerical data? Does it show the geographical and "built" features of a region?
2. For each type, create a sketch of that visual aid. Or, look through old magazines, junk mail, or old files for examples of each visual aid. These examples can show different, effective designs for each visual aid.
3. For each type, describe in writing a type of situation where this visual aid might be useful.

Chapter 9

Using Document Design Strategies to Create Usable and Appealing Documents

To become a good technical writer, you must know that successful documents are both well written and effectively designed. Understanding the elements of document design will help you to create documents that are usable and visually appealing.

Once technical writers reach the final stages of the document planning process, they understand a great deal about the document they will write. At this stage, technical writers understand the document's communication context, know what organizing strategy to use, and have identified places to locate more information about the topic. Visual aids have been chosen, and their placement within the document's outline has been considered.

The final stage in the planning process is document design. As a technical writer you need to know the answers to these questions about designing usable and appealing documents:

- What *strategies* related to document design do I need to know?
- What are the important *spatial elements* of document design?
- What are the important *textual elements* of document design?

Understanding how to design documents that are both useful and visually appealing is the mark of a good technical writer, and the most successful documents are both effectively written and well designed.

What strategies related to document design do I need to know?

Understanding three strategies that are related to document design may help you to decide what spatial and textual design elements to incorporate into your document and how to use these design elements most effectively.

The three strategies you need to apply include (a) identifying the document's communication context, (b) knowing your genre, and (c) chunking and labeling information.

Identifying the communication context

In Chapter 5 we discussed the importance of identifying a document's communication context—the situation, audience, and purpose of a document—before beginning to write. Identifying a document's communication context also helps you to decide what design elements to incorporate. Successful technical writers always identify a document's communication context before making design decisions.

Read through the two examples that follow. In both cases you can see that identifying the communication context of each document changes the design elements that are used:

> *Brochure for parents:* As a technical writer for a large hospital, you are writing an informative brochure about the benefits of good nutrition for pre-adolescent boys and girls. This brochure will be distributed to parents at the hospital. You design this brochure with clearly illustrated graphs and charts, bulleted lists noting information such as major health risks, and a professional-looking font style.

> *Brochure for children:* As a technical writer for a large hospital, you are writing a second informative brochure about the benefits of good nutrition, but this time the brochure's readers are the pre-adolescent girls and boys themselves. This brochure will be distributed to children at local schools. You design this three-color brochure with bright, attractive visual aids illustrating the various food groups, a great deal of white space, and a font style that resembles cursive handwriting.

In these examples, document design decisions were influenced by each brochure's communication context. The brochure written for parents has a style that is professional and exudes confidence in the information that is communicated. The brochure written for children has a style that is attractive, fun, and easy to read. Knowing the communication context is useful not only for planning the written, or textual, elements of a document but also for planning design elements.

Knowing the genre

Knowing what *type* of document, or genre, is appropriate for a specific communication context also helps you to decide what design elements to incorporate into that document. Typically, genres have certain characteristic design elements appropriate for different contexts. (See Chapter 3 for several different genres that technical writers typically create.)

Read through the example that follows. In this case you can see that knowing the appropriate genre affects the design elements that are incorporated:

> *Document for the user:* As a technical writer for a manufacturing company, you must write a document that describes how to use the important features of a new kitchen appliance your company produces. The readers for the document are those who will use the appliance.
>
> You decide that the most appropriate genre for this type of communication context is a user's manual. Certain design elements are characteristic of most manuals: illustrations, numbered lists of steps, and ample white space to aid in readability. Since most readers expect to find and need to use these design elements in a manual, you decide to incorporate these elements into yours.

Knowing what genre is most appropriate for this communication context allows you to choose certain design elements based on their importance to the genre. By using genre characteristics that are appropriate for the communication context, your document design decisions will be sure to result in a document geared to the intended audience.

Chunking and labeling

Chunking and labeling information in your document also helps you to decide what design elements to incorporate. Chunking and labeling not only help you to organize topics of information, but they help you to visually organize text on the page so that it is accessible to users. We discuss both the chunking and labeling processes in this section since writers often perform them at the same time.

Chunking is the process of grouping related information together, and *labeling* is the process of providing titles for each chunked group of information. Typically, information is chunked either by topic or according to audience use. The following example illustrates how information is chunked by topic.

Figure 9.1 shows the variety of topics selected to appear in the new *Employee Manual* for Shermann Financial Group, a large investment firm. Eventually, the manual will be given to all employees to familiarize them with the firm's policies and procedures.

Figure 9.1 A variety of topics for an employee manual

Employee topics
Purpose for the dress code
Paycheck regulations
IRA/retirement benefits information
Employee dress code regulations
Christmas account information
Employee penalties for disobeying dress code
Dress code exceptions

Figures 9.2a and 9.2b show how the writers decided to chunk and label each selection of the manual. Notice how much easier it is to recognize the appropriate topics when they are categorized and labeled clearly.

Figure 9.2a Employee dress code

A. Purpose for the dress code
B. Employee dress code regulations
C. Employee penalties for disobeying dress code
D. Dress code expectations

Figure 9.2b Employee salary information

A. Paycheck regulations
B. IRA/retirement benefits information
C. Christmas account information

Figure 9.3 illustrates how information is chunked according to audience use. This outline lists important topics that will be covered in a soil report for a local farmer. This report is used by the farmer to make informed decisions about ways to improve the soil, fertilizer, and seed used in each of the seven fields that he farms. You know that the farmer typically visits each field separately and that he needs information about all three topics per field.

Figure 9.3 A variety of topics for a farmer's field report

- Seed recommendations for Field 1
- Fertilizer recommendations for Field 3
- Seed recommendations for Field 3
- Soil recommendations for Field 1
- Fertilizer recommendations for Field 1
- Seed recommendations for Field 2
- Soil recommendations for Field 3
- Fertilizer recommendations for Field 2
- Soil recommendations for Field 2

Chunking this information according to audience use means that rather than chunking the information by topic (soil, fertilizer, and seed), you must chunk all of this information by field (Figures 9.4a–c).

Figure 9.4a Field #1

A. Soil recommendations for Field 1
B. Seed recommendations for Field 1
C. Fertilizer recommendations for Field 1

Figure 9.4b Field #2

A. Soil recommendations for Field 2
B. Seed recommendations for Field 2
C. Fertilizer recommendations for Field 2

Figure 9.4c Field #3

A. Soil recommendations for Field 3
B. Seed recommendations for Field 3
C. Fertilizer recommendations for Field 3

Knowing a document's communication context, its appropriate genre, and the ways you can successfully chunk and label information all make the document and page design decision process much easier and more effective.

Response Exercise 9.1

Chunking and labeling are important strategies related to designing usable and appealing documents. To understand how useful these strategies can be, find a user's manual for a product that you own, read through it, and write down your responses to the following questions about how chunking and labeling is used in the manual:

1. Write down the titles of the main sections of this user's manual. Are these examples of chunking and labeling?
2. Select one section of the manual, and examine it for chunking and labeling. How is information organized in this section? List the chunks of information that are discussed. Are these chunks arranged by audience use or topic?
3. Can you think of a more effective way to chunk the information either in the section you have selected or another section of the manual? If so, create an outline of the way you would chunk and label the information differently.

What are the important spatial elements of document design?

Spatial elements are those design features that manipulate the relationship between space and the text of a document. Primarily, spatial elements manipulate those places in a document that include the area near the edges of the page, between the lines of text, and surrounding the visual aids.

Technical writers manipulate space to ensure that their documents are as appealing and usable as possible. The most effective way that writers manipulate space is by choosing specific spatial design elements and using them to produce certain effects in a document. As a technical writer, you will use spatial elements differently for each document that you create.

Several important spatial design elements are described in this section. Understanding the types and uses of spatial design elements will enable you to create effective technical documents. Here we discuss five spatial design elements:

- Margins
- White space
- Visual aids placement
- Line spacing and leading
- Indenting

Margins

Margins, typically measured in inches, are the amount of space between the outside edge of the page and the point where the text begins. There are margins for the top and bottom, and left and right sides of your document. Many times, the margin widths will be different for each of these sides (Figure 9.5).

Figure 9.5 Differing margin widths of a block style business letter

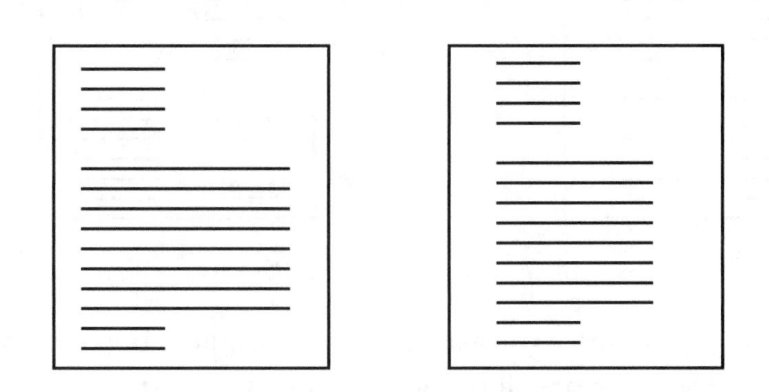

White space

White space constitutes those areas of your document that are absent of text or visual aids. For example, the space from the edge of the page to the beginning of the text in Figure 9.5 is white space. Manipulating white space in a document is very important because this process can affect readability. See Figure 9.6 for several different ways to balance white space and text on the page and thereby increase readability.

Figure 9.6 Ways to manipulate white space in documents

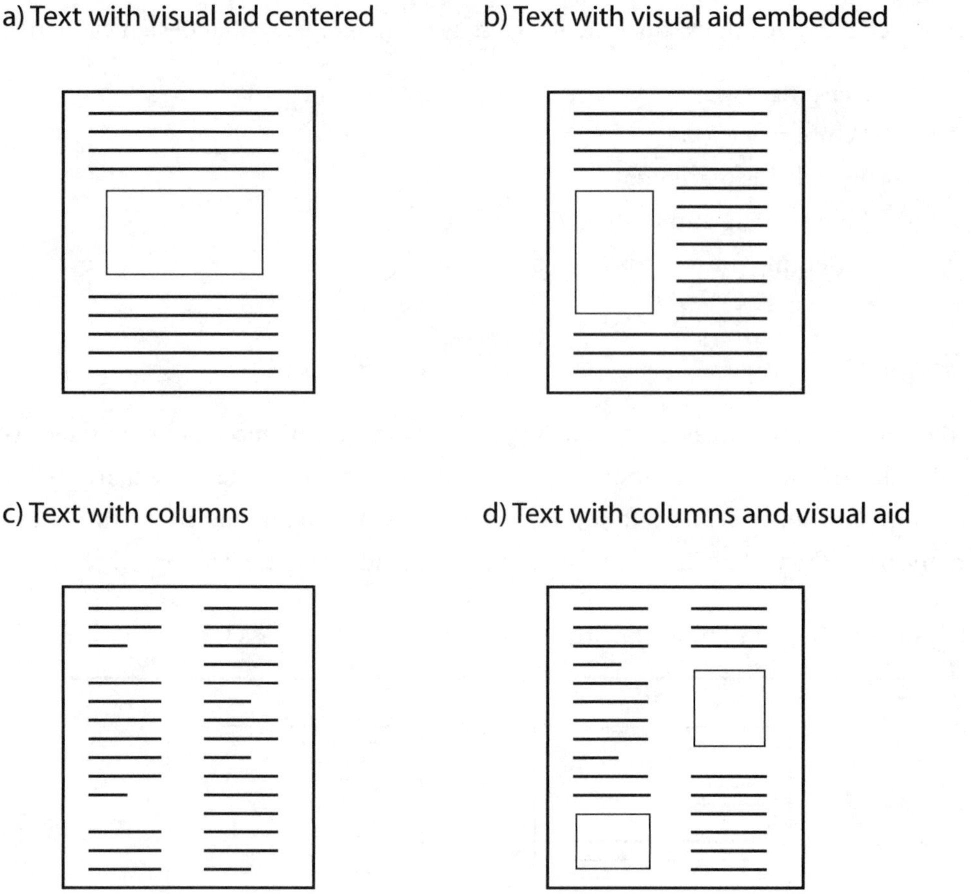

a) Text with visual aid centered

b) Text with visual aid embedded

c) Text with columns

d) Text with columns and visual aid

The amount of white space you allow within a document is important since too little hinders effective readability and too much wastes space. Be aware of the amount of white space you use in the margins, between text, and around visual aids.

Visual aids placement

The placement of visual aids on the page affects readability and is an important element in document design. Listed below are several strategies that successful technical writers use to effectively place visual aids within the documents that they create:

Questions writers ask themselves about *visual aids placement*

- Is the visual aid placed in proximity to the text that refers to it, usually following it closely?
- Is there at least $^1/_4$–$^1/_2$″ of white space surrounding the visual aid?
- Is the visual aid no larger than half of one page? (If larger, consider moving it to an appendix or reducing its size.)
- Is the visual aid balanced with the text that surrounds it?

Line spacing and leading

Line spacing, or leading, is the amount of full line spaces that you place between lines of text. For example, a document can be single spaced, double spaced, or triple spaced, depending on the document's use (Figure 9.7).

Figure 9.7 Single, double, and triple line spacing

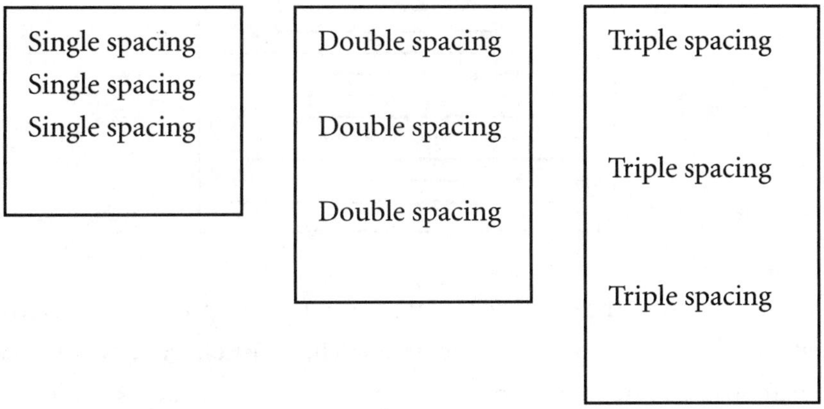

Single spacing	Double spacing	Triple spacing
Single spacing Single spacing Single spacing	Double spacing Double spacing Double spacing	Triple spacing Triple spacing Triple spacing

Leading is based on the same concept as line spacing, but is the miniscule number of white spaces between lines of text.

Indenting

Indenting creates more white space between the text and the left and/or right margins of a document. Indenting only changes margin widths within a portion of the document. As in Figure 9.8, indenting creates white space in an area where it is needed the most.

Figure 9.8 Indenting portions of text in a document

Spatial elements are those design features that manipulate the relationship between white space and text in a document. The most effective way that writers manipulate space is by choosing particular spatial design elements and using them in specific and purposeful ways. Technical writers use spatial design elements to ensure that their documents are as appealing and usable as possible.

Now that you understand the different spatial design elements from which you may choose, you must be able to identify the textual design elements that you may use as well. In the next section we identify the most important textual design elements and describe how to use them effectively in your documents.

Response Exercise 9.2

Knowing how to choose and use the five spatial design elements you learned about in this section will help you to design more usable and appealing documents.

Select at least two documents that you have access to: a letter, user's manual, instruction booklet, information brochure, or any other documents you can locate. Read through each document and write down the spatial elements used in each. Next, respond to the following questions:

1. Is each element used? If not, why? Should each be?
2. In writing, describe how effectively each element is used in the document.
3. In what ways would you change the way each element is used? Why?

What are the important textual elements of document design?

Textual elements are features in a document that manipulate text. These elements are used for organizing or emphasizing text. Here are the most important page design elements:

- Headings and subheadings
- Font style and size
- Format devices
  - All caps, boldfacing, italics, underlining
  - Bullets, numbered lists
  - Color, shading, boxes

Headings and subheadings

Headings are the titles of each important section of your document. Headings may be one word or several words, and they can be nouns, noun phrases, or questions. Subheadings are the titles of each subsection; like headings, subheadings can be nouns, noun phrases, or questions that are one word or several words in length.

Listed below are several questions that successful technical writers ask themselves about headings and subheadings.

Questions writers ask themselves about *headings* and *subheadings*

- Is there adequate white space between the heading (subheading) and the surrounding text?
- Are all headings formatted consistently?
- Are all subheadings formatted consistently?
- Are the headings and subheadings formatted differently than one another?

- Are the headings and subheadings exhibiting parallel structure? In other words, are they relatively the same word length and is each constructed similarly (either all noun phrases, words, or questions)?

Font style and size

The *font style*, or typeface, refers to style of the characters. Because of the improvements in software programs, technical writers can choose from a range of different font styles.

As a writer, the font style that you use may be the style that your company favors. For example, many companies prefer the professional look and easy readability of a Times font style, while other companies may favor a different font style such as Courier or Arial.

There is a wide variety of font styles available, and if you are free to choose the font style, keep these questions in mind:

Questions writers ask themselves about *font style*

- Is this font style readable and professional looking?
- Is this font style appropriate for the document's communication context?
- Is this font style easy to copy or reproduce?

Another point to consider about font style is whether the font style is a *serif* or a *sans serif* type. The serif type of font style has tails attached to the letters, while the sans serif type is free of these tails (Figure 9.9). Generally, the serif font is easier to read on the page, while the sans serif font style is easier to read on a computer screen.

Figure 9.9 A serif and sans serif style

This is a serif font style.

This is a sans serif font style.

The *font size* refers to the number of points or pixels per inch in each letter. Once again, because of the improvements in software programs, technical writers can choose from a range of different font size options. Figure 9.10 illustrates a sample of the different font sizes from which writers can choose.

Figure 9.10 Font size choices

9 point 12 point 14 point 16 point 18 point 24 point

Format devices

A number of different format devices organize or add emphasis to text. Below is a list of the most useful format devices, followed by a description of each device:

- All caps, boldfacing, italics, underlining
- Bullets, numbered list
- Boxes, color

All caps

The *all caps* format device adds emphasis by placing text, either one word or entire sentences, into all capitalized letters. The all caps function, like most format devices that add emphasis, must be used sparingly and purposefully. Notice in Figure 9.11 that the more you use this function (an entire sentence rather than one word), the text that is in all caps becomes more difficult to read. Therefore, avoid all caps text for more than short headings, and do not use all caps text for sentences.

Figure 9.11 Showing all caps readability

This single word has useful emphasis to capture the reader's attention:

CAUTION

This sentence has too much all caps text that is difficult to read, which diminishes effectiveness:

PRESS THE SAFETY BUTTON AFTER EACH OPERATION; IF YOU DO NOT PERFORM THIS OPERATION, THE MACHINE MAY SUFFER DAMAGE.

Boldface

The *boldface* format device adds emphasis to text by making the font style heavier looking and darker than normal. Like the all caps function, it is important to use boldfacing consistently, sparingly, and purposefully.

Notice how the phrase that is boldfaced in Figure 9.12 is quite easy to distinguish from the surrounding text.

Figure 9.12 How boldfacing adds emphasis

> This text is an important one for four reasons, and the first and most important reason is **Reason A**. This reason is important because it uses all the functions of the other three reasons but does so more efficiently than ever before.

Italics

The *italics* format device adds emphasis by making text appear scripted. As with the all caps and boldfacing functions, it is important to use italics consistently, sparingly, and purposefully.

Notice in Figure 9.13 how the italics function helps to differentiate between the words being defined and the definitions themselves.

Figure 9.13 How italics add emphasis

> *Hardware:* Those pieces of equipment or tools that help our computer system function.
> *Software:* Those programs that enable our hardware, or our tools and equipment, to run efficiently.
> *RAM:* Random Access Memory is the memory that your computer uses most often.

Underlining

The *underlining* format device adds emphasis by placing a line underneath one word or an entire sentence. As with other format devices that add emphasis, use underlining consistently, sparingly, and purposefully. Notice in Figure 9.14 how underlining is used to add emphasis to the phrase.

Figure 9.14 How underlining adds emphasis

It is important to remember that all of the functions will be turned down at <u>5 p.m. on Friday, November 30,</u> and that there will be no exceptions to this rule.

Bullets

The *bullets* function is a format device that lists items in text and places a bullet—a circular symbol—in front of each item in that list. Bullets are useful for adding emphasis to a list of items.

Notice the difference in format between the bulleted list (Figure 9.15a) and the list with asterisks placed in front of each item (Figure 9.15b). The items in the bulleted list are more professional looking and exude a crisper, formatted look.

Figure 9.15a Bulleted list

The following is a list of all of the approved colors to be used in the Homecoming floats:

- Green
- Blue
- Red
- Yellow

Figure 9.15b Asterisks list

The following is a list of all of the approved colors to be used in the Homecoming floats:

* Green
* Blue
* Red
* Yellow

Numbered list

The *numbered list* function is a format device that lists items in text and places a number in front of each. Numbered lists are similar to bulleted lists in that they add emphasis to a list of items. However, one important difference between them is that numbered lists suggest a hierarchy or a step-by-step method, while the bulleted list suggests that the items are of relatively equal value and importance. Therefore, use numbered lists to suggest order (Figure 9.16), and use bulleted lists to suggest equal ranking.

Figure 9.16 Numbered list adds emphasis and suggests order

1. Dial the system number, 555-1234, using a touch tone phone.
2. After the "welcome message," press 1.
3. Listen to the list of options. Choose your option and press the # key.
4. Listen to your selected option, and wait for the operator.
5. To replay your list of options, press 9.

Boxes

Placing text or a visual aid inside a *box* is a useful way both to add emphasis to that information and to distinguish it as different from other information. The "Connie's story" elements from Chapter 5 are examples of these boxed items.

Color

The use of *color* within any document will naturally add emphasis to the information that it is highlighting, since color on the page is the first element that readers notice. Remember, documents that use color are more costly to reproduce, so use color sparingly.

Listed below are some questions that technical writers ask themselves about using color in documents:

Questions writers ask themselves about *color*

- Is color within my budget for this project?
- Is color used consistently throughout the document? For example, if I use green to denote sales increases in the first section of the document, did I continue to use this color for sales increases throughout?
- Is color used sparingly throughout the document? For example, have I limited my use of color to three colors, such as red, green, and blue?

Response Exercise 9.3

Knowing how to choose and use the textual design elements you learned about in this section will help you to design more usable and appealing documents.

Select at least two documents that you have access to: a letter, user's manual, instruction booklet, information brochure, or any other documents you can locate.

Read through each document and write down the textual elements used in each. Next, respond to the following questions:

1. Is each of the elements used? If not, why? Should each be?
2. In writing, describe how effectively each element is used in the document.
3. In what ways would you change the way each element is used? Why?

Technical writers understand a great deal about the document they will write by the time they reach the final stages of the document planning process. At this final stage is document design. A successful technical writer understands how to design effective documents that are both useful and visually appealing.

Chapter 9 Task Exercise

Scenario

Choose an interesting article from the op-ed page of the newspaper. Select an article that not only interests you, but also includes at least one visual aid. Use the information from the article to create a document that will be read by an audience of your peers. They will read the document to learn more about the topic being discussed.

Not only do you want the document to be useful, you want it to be visually appealing as well. How would you take the information from this article and redesign it into a usable and appealing document that could be easily scanned and read by an audience of your peers?

Task

On a separate sheet of paper create a sketch of the document. Be sure to think about and use the design strategies you learned about in this chapter. Also, select those spatial and textual design elements that you believe would be most useful for this document's communication context.

After you have created a sketch of the document and its design elements, respond to the following questions about the choices you made.

1. What design strategies did you use? How was using these strategies useful?
2. What spatial design elements did you select? How did you select the most appropriate ones for the document's communication context?
3. What textual design elements did you select? How did you select the most appropriate ones for the document's communication context?

Part III

Using Effective Processes

Chapter 10

Collaborating With Colleagues and Writing Effectively Within a Group

To become a good technical writer, you not only need to write effective documents on your own, but you also need to know how to write documents collaboratively with your colleagues.

Sometimes documents are assigned to teams rather than individuals. As a technical writer, you often have the responsibility of planning and drafting documents by yourself; however, a majority of technical writing that you do on the job will be done collaboratively. This means that you and your colleagues will share the responsibility of producing a quality document or a set of documents.

You need to ask three important questions when collaborating with colleagues. Answering these questions will help the collaborative process go more smoothly and allow your team to produce a more effective document:

- In what *situations and with whom* will I collaborate?
- What *strategies* can I use to be an effective collaborator?
- How can I use *conflict* to make collaboration more effective?

Working with colleagues to produce a quality document can be a very rewarding experience. Not only are you given the opportunity to initiate or deepen your professional relationship your colleagues, but you have the opportunity to improve as a writer by observing how others plan and draft documents.

In what situations and with whom will I collaborate?

Many writing situations call for collaboration; as a technical writer, you will be involved in many projects in which documents are prepared by a team. While your own workplace dictates when and how frequently you collaborate with others on planning and writing a document or set of documents, two collaborative situations and types of collaborators are typical.

Collaborative situations

As a technical writer, you may be involved in two basic types of collaborative situations: (1) you collaborate with your team on all aspects of the writing process—from document planning through revision, or (2) you consult with one or more of your colleagues to draft specific sections of a document.

For example, in order to write an accurate and effective hardware description in a recommendation report you might work with the engineer who designed the hardware. She may help you draft the hardware description, but you draft the remaining sections of the recommendation report on your own, following the general information the engineer provides. Or, you might consult with other colleagues during a particular phase of the writing process. For example, during the final stage of writing an informational brochure, you might ask a colleague to read through and edit the brochure for accuracy, design consistency, and mechanical correctness.

As a technical writer, you will be involved in many writing situations that benefit from the collaboration of others. Since collaboration in the workplace is inevitable, keep in mind the many benefits to working with a team.

Your benefits

You receive many benefits as a result of collaborating with colleagues on a project. While collaboration is a time-consuming and labor-intensive process, you will discover that spending quality time with your peers on a project that matters will help you to *cultivate or strengthen your professional relationships* with colleagues.

When you collaborate with others on a project, you observe how your colleagues plan and draft documents. One hallmark of a successful writer is a willingness to discover new ways to improve the writing process. Collaboration helps you to *learn more about yourself as a writer* by observing how others approach the writing process and how they use written documents to solve communication problems.

As a technical writer, you will probably collaborate with a variety of people over the course of your career. Your collaborators may work in a variety of areas within your organization: engineering, graphic design, publication, communication, or marketing. These colleagues bring with them relevant issues, ideas, and challenges from their own fields to your team. Not only will these elements impact the document that your team creates, but as a writer you will be influenced by them as well. Therefore, collaboration helps you *to learn more about different fields.*

Document benefits

Each team member receives several benefits as a result of collaboration. However, the document that you and your team create also benefits from collaboration. A document that is written using the combined efforts of a team of people who have different perspectives and a variety of talents and interests is a richer and more effective document than one produced by an individual.

A team of colleagues who collaborate on a document may also devote more energy and time to planning, drafting, and revising that document than would be possible with an individual writer. Simply put, a document written by several different people has more work put into it.

As a technical writer, you will be involved in many collaborative projects, so you need to know what situations and with whom you will be likely to collaborate. You can identify many benefits of working on a team written or colleague consulted document. In the next section, we discuss several strategies of effective collaboration.

Response Exercise 10.1

Collaborating on a team written or colleague consulted document has many benefits. To identify the variety of benefits that collaborating can provide, write down your responses to the following questions:

1. If collaborating on a document is more labor- and time-intensive than writing that same document individually, how do I benefit from producing a collaboratively written document?
2. What types of colleagues could I collaborate with? What would be useful about collaborating with people from different fields?
3. In what ways could a team written document be more effective than a document generated by an individual?

What strategies can I use to be an effective collaborator?

Regardless of your collaborative situation, you need to keep several strategies in mind to help you be an effective collaborator. These strategies are useful for both team written and colleague consulted documents.

The strategies we discuss in this section are grouped into three categories: preparation, team meeting, and follow-up. These categories reflect a typical collaborative process in that most collaboration in the workplace is organized around one or more team meetings.

The *preparation* category shows how to get ready for each meeting while the *team meeting* category lists strategies to implement during the meeting itself. The *follow-up* category notes how you can be a productive team member after a meeting.

Preparation	Team meeting	Follow-up
• Perform assigned tasks	• Show enthusiasm	• Reflect
• Be prompt	• Be cooperative	• Communicate with team
• Create a meeting agenda	• Ask questions	• Communicate with others
	• Listen	

Preparation

Both you and your team benefit if you follow these strategies regarding team meeting preparation. These strategies can be used to prepare for all of your team meetings.

Perform assigned tasks

Typically, team members are delegated specific tasks to complete by a certain deadline, and these tasks are usually delegated according to expertise and interest. Finishing tasks that are delegated to you shows your collaborators that you are a conscientious professional and helps to build trust among team members.

Be prompt

Team meetings proceed much more smoothly when meetings are scheduled on days and times that are possible for all members. When no one has to show up late or leave early, meetings are sure to be productive.

Create a meeting agenda

A team meeting with a specific meeting agenda is more likely to accomplish team goals. You and your colleagues are more likely to stay on task and be satisfied with your group's progress if an agenda is followed.

Assign one team member the task of creating a meeting agenda, which lists objectives and tasks for the next meeting, and is distributed to all members before that meeting. That way, you and your colleagues know not only what will be discussed, but also how to prepare for that discussion.

Team meeting

You gain several benefits by understanding and following these team meeting strategies. Like the meeting preparation strategies, these strategies can be used for any meeting that you and your team schedule.

Show enthusiasm

Sound too obvious? It's not! Some people mistakenly believe that being neutral, being cool is the same as being professional. True, some team projects will be more interesting to you than other projects, but you need to find something about each project that really interests you. Identifying a point of interest in the project enables you to be enthusiastic about the work your team is doing. During team meetings you need to express this enthusiasm to your collaborators. Since enthusiasm and a positive attitude are infectious, your team members will probably respond positively as well.

Be cooperative

A collaborative project is actually more time- and labor-intensive than a project you work on alone. However, the benefits of collaboration are many, and one way to get the most out of collaboration and particularly out of team meetings is to be cooperative. Cooperate with your team members and be willing to try new things and perform new tasks.

Ask questions

As a collaborator on a team, you have a responsibility to ask questions of your team members in order to understand clearly the topics they are discussing. Not only should you prepare your own topics for discussion during the meeting, but you should make sure that you understand your team members' discussions as well.

In addition to asking questions about content, you need to ask questions about the purpose and audience of the document, about how to explain and organize the information, and about ways to illustrate and design the document.

Listen

One crucial aspect of being an effective collaborator is knowing how to listen. Make sure that you hear your collaborators—and more than that, make sure that you actually listen to them. You can keep track of what you listen to by taking careful notes. These notes, or *meeting minutes*, are valuable resources for both you and your fellow team members. Listed below are several questions that writers ask themselves about taking meeting minutes:

Questions writers ask themselves about *meeting minutes*

- Have I identified the team members who are present? Also, have I clearly noted the meeting's purpose and dated the meeting minutes?
- Have I identified and discussed each important point raised by our team? Have I described this discussion completely enough so that people who did not attend the meeting will understand what occurred?
- Have I indicated the actions or tasks resulting from that discussion?
- Have I created a consistent organizational format for the meeting minutes so they are easy to scan and understand?

Often, a team will assign one person to take the minutes for each team meeting. If a team member is assigned this task, be sure that you receive a copy of the minutes for each meeting. Also, you may choose to take your own notes during the meeting minutes in addition to your team's meeting minutes. One of the easiest ways to distribute meeting minutes is to send them to team members by e-mail.

Follow-up

Your responsibility as a productive collaborator does not end once your team meeting concludes. You must be aware of the usefulness of meeting follow-up strategies that enable you to get the most out of your team's collaboration through reflection and communication.

Reflect

Immediately following your team meetings, you should find ten minutes or so to reflect on the topics raised, the points discussed, and the actions taken as a result of your team's collaboration. Reflecting on these elements forces you to digest all aspects of your team's interaction; as a result, you will more easily remember and benefit from the meeting and your collaboration.

Another useful way to reflect on your collaboration during a team meeting is to reread the meeting minutes. (If you are assigned to prepare the minutes, you could take this time to type up the minutes.) Then, if you have you have further questions for your colleagues, you can contact them immediately. This reflection is also a good time to begin preparing the agenda for your team's next meeting.

Communicate with team

For a team to be successful, it must have effective and efficient communication among its group members. Members can contact one another in a variety of ways: face to face, telephone, print memos, or e-mail. One aspect of being a good collaborator is to identify the most efficient methods for your team members to communicate with you.

Communicate with others

Team members not only must communicate with each other, but they also must communicate with others outside of the group. Supervisors, clients, or other colleagues may need to keep in contact with your group to chart your team's progress, offer suggestions, or provide recommendations. As a team, decide how to keep these important lines of communication open and determine who represents your team as the contact person.

Regardless of your situation, keeping these strategies in mind while you are a member of a team will enable you to become a more effective collaborator. Not only will you become a more productive writer, but the teams to which you belong may become more successful as well. These strategies—from preparation, team meeting, to follow-up—are useful for improving your individual performance and your team's performance as well.

Response Exercise 10.2

Technical writers use three important types of strategies to improve their collaboration skills. Keeping these strategies in mind, read the following scenario and respond to the questions:

Scenario

As a technical writer for a large insurance firm, you have been asked to collaborate on an informational brochure about the new policy services and standards your firm will implement. This brochure will be distributed to all current customers as a way to inform them about these new services and standards. You will be collaborating with two insurance underwriters on this project. The first planning meeting is scheduled in two days. Write down your answers to the following questions about this and subsequent meetings:

1. Name two ways that you can prepare for your first team meeting. What will you do differently to prepare for the next team meeting?
2. Think about what topics you and your team members may discuss during your first planning meeting, and create a list of these possible topics.
3. What types of strategies can you use during this team meeting to be an effective collaborator?

4. Imagine that your first team meeting was successful, and each member left with a task to perform. What strategies can you use after this meeting to be a productive team member?

How can I use conflict to make collaboration more effective?

All too often, collaborators believe that the best teamwork is conflict free. However, successful teamwork depends on individual members knowing how to use the right kind of conflict to help make collaboration as successful as possible.

In this section you will learn about three different types of conflict that commonly occur in workplace collaboration situations: procedural, affective, and substantive conflict (Burnett 1997, 103). Then you will discover how conflict can be used to make your team an even more successful and productive one.

Procedural conflict

Procedural conflict involves disagreements about factors such as when to meet, how meetings should be run, who should be in charge (if anyone), and how decisions should be made. While this kind of conflict can be destructive, you can usually prevent it from happening by careful preliminary decisions.

Any collaborative group or team should make dealing with procedural decisions one of its first items of business. Specifically, team members should openly discuss and come to agree about the following questions.

Questions writers ask themselves about *procedural conflict*

- Do my group members agree about when, where, and how long we meet?
- Do my group members agree about the team's overall goal?
- Do my group members agree about each person's role in accomplishing that goal?
- Do my group members agree about how decisions will be made and how disagreements will be resolved?

Affective conflict

Affective conflict involves interpersonal disagreements that involve your personality, values, attitudes, and biases. This kind of conflict can be very destructive if you do not recognize it in yourself and in others.

One way to combat affective conflict is to concentrate on making your team's product—the written document—as successful as possible. Rather than focusing on other team members and your attitudes towards them, try to focus on the document that your team is creating.

Another way to combat affective conflict is to identify your own prejudices and biases, and then make a conscious decision to leave them at the door when you are working with collaborators. Any prejudices you have about gender, race, nationality, ethnicity, sexual orientation, age, religion, body type, or personality type—just to mention some unfortunately common areas for unwarranted prejudice—can interfere with your team's efforts.

Questions writers ask themselves about *affective conflict*

- What are my personal biases or prejudices?
- Am I willing to leave my personal biases or prejudices out of my collaborative interactions?
- Am I willing to privately confront another member of my team (and suggest alternative, more productive behavior) if I see his or her biases or prejudices interfering with the team's work?
- Should I overlook another member of my team even if I see his or her biases or prejudices interfering with the team's work?

By shifting your focus and energy away from your personal prejudices and toward the document, you will be able to engage more readily in a productive type of conflict: substantive conflict.

Substantive conflict

Substantive conflict involves the decisions your team makes about the elements included in the document. By engaging in substantive conflict—or thinking critically about your document's communication context, content, organization, prose style, and design—your team carefully considers a *variety* of different ways to approach the document. Substantive conflict helps team members consider a range of ideas about how to produce a better document, and this type of conflict usually helps your team to create a better product. One of the best ways to make substantive conflict work for your team is for all members to be aware of the benefits of this type of conflict.

Questions writers ask themselves about *substantive conflict*

- Do my group members understand the difference between procedural, affective, and substantive conflict?
- Do I promote substantive conflict within my team's document planning or drafting sessions by asking critical questions? Whenever necessary, do I play "devil's advocate" to prompt critical thinking?

- Do I help my team stay on task by making the document the focus of our discussion?
- Do I accept when others engage me in substantive conflict? Do I respond by building on the ideas of other team members? Do I allow (even encourage) my team members to challenge my decisions about the document?

As a writer and collaborator, remember that you should be able to recognize the differences between procedural, affective, and substantive conflict. Engaging in substantive conflict makes the document your team produces the very best one possible.

Response Exercise 10.3

Successful collaborators can often identify three types of conflict. One type is much more productive than the others. Keeping these three types of conflict in mind, reread the following scenario and respond to the questions.

Scenario

As a technical writer for a large insurance firm, you have been asked to collaborate on an informational brochure about the new policy services and standards your firm will implement. This brochure will be distributed to all current customers as a way to inform them about these new services and standards. You will be collaborating with two insurance underwriters on this project.

Presently, you are attending your first planning meeting. Write down your responses to the following questions:

1. When your team first gets together, one of your team members says he's brought a draft of a brochure that should meet with everyone's approval, so the team can make short work of the project. Another suggests that everyone should have input and that the team should take time to discuss its schedule and ways to make decisions. What kind of conflict is at work here? What do you do?
2. One member of your team begins to discuss office politics. You disagree with his position on one issue, but this issue does not directly impact the brochure your team has been asked to create. What kind of conflict is at issue here? What do you do?
3. During the discussion your team is having about the order of points for the brochure, one of your team members suggests a possible organization of these points. You believe part of her idea might work, but it needs revision; specifically, you believe that there is a way to reorganize her outline so the audience will better understand the brochure's message. What kind of conflict is at work here? What do you do?

4. As your team meets for the third time, you see one of the team members lean to the next person, and you overhear that person whisper a sexually explicit joke about another member of the team. While they laugh quietly, you are angry and embarrassed. What kind of conflict is at issue here? What do you do?
5. One of your team members disagrees with your brochure design ideas, and he wants to explain to the group his ideas about other brochure design possibilities. What kind of conflict is at issue here? What do you do?

Working with others, whether these colleagues are engineers, designers, supervisors, or fellow writers, can be a very rewarding professional experience and can result in the creation of a high quality document. To make any collaborative writing experience more productive and successful, you need to remember how you can improve as a collaborator: prepare for team meetings, perform well during meetings, and reflect and communicate after meetings.

Chapter 10 Task Exercise

Scenario

Imagine that you work as a technical writer at the home office of a hardware megastore outlet that sells home improvement supplies. Your customers include homeowners, do-it-yourselfers, and contractors. The home office needs you to work with a team of managers, supervisors, and legal personnel to research a possible new store site.

You have been asked to collaborate with this team to research three possible outlet sites, recommend one, and a write a report about your team's findings. This report must describe the three sites that your team has researched and clearly recommend one site. Your first team meeting, with all of your collaborators, is scheduled for next week.

Task

1. List and describe the strategies that you can use to prepare for this initial planning meeting, to perform effectively during the meeting, and to plan for your next team meeting.
2. You have been asked to write the meeting minutes for this initial planning meeting and the remainder of your team meetings. List the strategies that you will use to create effective meeting minutes. Next, write down at least two possible ways that you can communicate these meeting minutes to your group.
3. Your supervisor, who is a member of your team, has asked you to prepare a short presentation about the benefits of using conflict within collaboration. She has requested that you (a) identify and describe the different types of conflict that groups typically exhibit and (b) discuss the

usefulness of one type. Your supervisor has also asked that you (c) engage your colleagues in a role-playing activity in which they practice how to initiate positive conflict and how to respond to three types of conflict during team meetings. Devise a list of possible situations that your team members could act out that would help them engage in all types of conflict.

Chapter 11

Writing Effectively by Understanding Planning and Drafting

To become a good technical writer, you need to understand in what ways writing within an organization is unique and what is involved in planning and drafting a document.

Technical writers often find themselves writing an entire document with a team of two or more colleagues, and writers often have colleagues read through and suggest ways to improve a document. As a technical writer, you must not only understand *when* you need to collaborate with your colleagues on a writing project, but also *how* you move through the writing process (Figure 11.1).

Figure 11.1 Four stages of the writing process

Planning ←→ **Drafting** ←→ **Revising** ←→ **Editing**

- Identify communication context
- Brainstorm/locate more information
- Make document design decisions

Being a successful technical writer involves knowing how you work and write within a team of colleagues and knowing how your own writing process—such as how you perform planning and drafting—typically works.

In this chapter we identify and discuss what is unique about writing within an organization, and we explain how you can begin to understand your own writing process. To be a successful and productive technical writer, you must understand the answers to the following questions about the writing process:

- How does *writing within an organization* affect my writing process?
- What do I need to know about *planning?*
- What do I need to know about *drafting?*

Answering these questions not only allows you to identify the special constraints that technical writers work around because they write within organizations, but also enables you to identify how you plan and draft.

What is unique about writing within an organization?

When you work as a technical writer, the writing you do within an organization is much different than the writing you do at home or in school. Writing within an organization means that you not only must understand your own writing process, but you must be able to embed that writing process into the schedules, deadlines, and rules or constraints of an organization.

When you write at home, you may be writing for yourself; that is, the only schedule, deadline, or rules that you have to follow are those that you dictate. When you write at school, you may be writing for a limited audience—the instructor who may have a flexible schedule and suggest rules that apply in a narrow academic setting. In the workplace, however, more often than not, a whole host of people may be involved, in one way or another, in your writing process. In this section, you will learn what to expect and what strategies to use when writing within an organization.

Deadlines and time management

In the workplace, technical writers must learn to produce quality documents in a short period of time; therefore, as a writer you must know what your deadline is and how to meet it. When writers are given a long-term deadline for the document, they usually create for themselves multiple, short-term deadlines.

Technical writers know that deadlines affect how they plan the document and how long they are allowed to draft and revise. You understand what document deadlines are in place for your document and then work within those deadlines to produce a quality product. Once you are familiar with your own writing process—from the planning through the revising stages—you know approximately how much time it takes you to produce a usable and appealing document.

Therefore, once you understand how you typically approach a writing task and how long it takes you to produce a document, you can begin to implement time management strategies. Technical writers use several strategies when managing their time. Below are questions writers ask themselves about time management:

Questions writers ask themselves about *time management*

- What is my document deadline? How much time do I have to produce a quality document?
- Will I be collaborating with others to produce this document? If so, what are their schedules, and when will we begin planning/drafting/revising?
- What tasks are on my schedule now? Will my new document planning and drafting tasks take priority over other projects?
- How long will it take me to plan this document? How much time should I give myself to research this topic?
- How long will it take me to draft this document? How much time do I need to edit or user-test this document?

Like other business professionals, technical writers often find that planning their schedules using a simple day planner or scheduling software helps them to manage their time more effectively.

Shorter projects that involve producing briefer or less involved documents may only require one or two preliminary document deadlines, which you can pencil in on your calendar or write on a note for your bulletin board. More complicated projects, however—particularly those involving a team—may need more complicated methods of time management.

Figure 11.2 presents an example of one type of time management tool, a Gantt chart. This chart requires you to identify and label small, manageable tasks that are part of the overall project; assign each of these tasks a time frame and completion deadline; and input these tasks into the overall timeframe allowed for the project.

Figure 11.2 Team project Gantt chart (first seven days only)

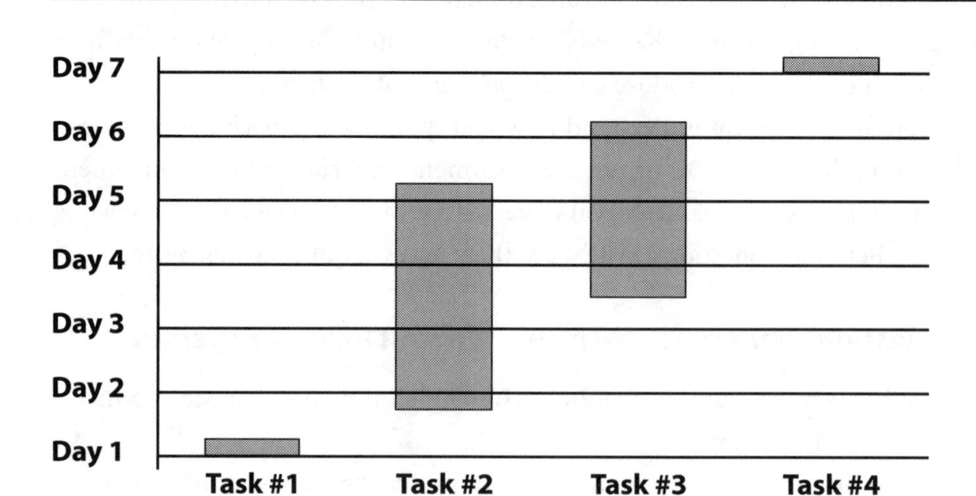

Gantt Chart Key

Task #1 First team meeting: identify team goals and individual tasks, schedule meetings
Task #2 Begin my individual task: search for topic information
Task #3 Write my topic information report
Task #4 Second team meeting: present my report

The Gantt chart allows you not only to identify specific tasks that you need to complete, but also to compare tasks. Therefore, you can see that one task is more involved and will take several days to complete while another task is less involved and may take only an hour or two to perform.

Remember, a Gantt chart is useful not only for planning your own writing schedule, but for planning around and with the schedules of others on your project team.

Response Exercise 11.1

As a technical writer working within an organization, you benefit from understanding and following time management strategies. To illustrate why time management strategies are important, write down your responses to the following questions:

1. Why do writers working with organizations benefit from effectively implementing time management strategies?
2. What are three important questions you should ask yourself about time management strategies? Why are these responses important to understand?

3. What methods could you use to help yourself manage and schedule your time?
4. In what ways do you currently use time management strategies? How could you better streamline these practices?

What do I need to know about planning?

In order to best use the Gantt chart or any other time management strategy, you must understand your own writing process. Typically, writers pass through a planning stage before they begin to seriously draft the document. Planning is essential for documents written both individually and with a team.

Writers pass through three phases during planning. Each phase is identified and described below. Successful technical writers ask themselves several questions as they move through these phases.

Communication context

The first phase that many writers pass through during planning is the phase in which they identify the document's communication context. In Chapter 5 we discussed three elements that are part of any document's communication context: situation, purpose, and audience. Identifying and understanding these elements are important steps in the invention and planning stages.

Writers ask themselves the following questions about a document's communication context. You may want to refer to Chapter 5 for more detailed information about identifying and understanding the three key elements of the communication context.

Questions writers ask themselves about a document's *communication context*

- What is my document's *situation?* What is my project schedule? Will I have collaborators? What is my prior knowledge about this document's topic?
- What is my document's *purpose?* What is this document's objective? What work will this document perform? What is this document's value?
- Who is my document's *audience*—are they customers, clients, supervisors? What previous knowledge does my audience have about this topic? What are their values and interests? How will my audience use this document?

Brainstorming

The second phase that many writers pass through during the planning stage is brainstorming. This is the phase in which writers invent and explore ideas about their document's topic. During brainstorming, writers may find that they need to locate more information about their topic. Locating information, a topic that is discussed at length in Chapter 7, may be a result of your brainstorming process.

Many times, beginning technical writers believe that they will not need to brainstorm ideas about the topics they write about, thinking that these topics will be defined for them. However, while technical writers may not brainstorm in all situations to the extent that a short-story writer may brainstorm, the process is still necessary.

Brainstorming may be defined in different ways. For some writers, it may be as simple as creating a list of major points or a developing a brief outline of important topics. For other writers, it means identifying an important topic and writing several paragraphs about their knowledge of that topic. For still others, it means devising a list of information they need to access in order to begin creating a substantial outline.

Listed below are several of the most typical questions writers ask themselves when they are engaged in brainstorming:

Questions writers ask themselves about *brainstorming*

- What is my document's topic? What do I know about this topic?
- What information should I include as major points?
- What information does my audience need and what information should I exclude?
- What is the argument I am making in this document? What audience type will I need to persuade? What strategies will I use to persuade them?
- What do I need to know about my topic? How can I access this information?
- What questions do I have about moving forward in this stage of invention and planning? Who could answer these questions?

Brainstorming is one of the most important parts of planning. You may return to the brainstorming phase throughout your document planning. Successful technical writers often find it necessary to brainstorm throughout the entire writing process. Brainstorming is a creative way to generate ideas for your document and to produce a usable and appealing document.

Organization and design

Another phase that many writers pass through during planning is organization and design. This is the phase in which writers begin to decide how they will organize their document's main points and how they will design the document to suit its communication context.

The organization phase is discussed in length in Chapter 6, which describes organizing and outlining documents, and document design is discussed in Chapter 9. Read through these chapters for more detailed discussions regarding these concepts. Remember, the organization strategy and the document design that you plan to use in your document must suit that document's communication context. Therefore, you must understand that context before you make organization and design decisions.

Listed below are several questions that writers ask themselves about organization strategies and document design:

Questions writers ask themselves about *organization and design*

- What are my document's main points? What organization strategy is most suitable for ordering and discussing these points?
- What examples or explanation should I provide for each of these points?
- What design strategy is suitable for my communication context?
- Does this design strategy complement my method for organizing the information?
- What information is best presented visually?
- What design strategy has been successful in the past with documents such as these?

As with the other two phases—identifying the communication context and brainstorming—the decisions you make about organization and design continue to change as you proceed through the planning stage of the writing process.

Response Exercise 11.2

As a technical writer working within an organization, you will benefit from identifying and understanding how you proceed through planning stages. To illustrate how you move through these stages, write down your responses to the following questions.

1. What three phases make up planning? Identify two important questions that you should ask yourself about each phase.
2. Why must you understand these three phases? Do all writers pass through the same phases?

3. Think about a document that you have recently written (it could be a paper for school, a letter to a friend, or a report or brochure written at your workplace). Describe in writing the invention and planning phases you went through to create that document. Are they different or similar to the phases described in this section? How could you make these phases more efficient or productive?

What do I need to know about drafting my document?

As a technical writer, much of your time will probably be spent moving through this important stage of the writing process: drafting your document. This is the stage in which you make important decisions about what information to include, what examples to provide, how to best communicate and argue your point, and how to persuade your audience.

Figure 11.3 Four stages of the writing process

Planning ↔ Drafting ↔ Revising ↔ Editing

- May use outline or notes to help generate text
- Continually make communication context, content, organization, prose style, and design decisions
- May collaborate with others to help generate text

Drafting is a different process for everyone, and the way that you pass through this stage is probably different than the way other writers pass through this stage. For example, some writers use an outline and do the majority of their drafting on a computer. Other writers use a computer only to draft the final stages of a document, after the bulk of the planning has been done with paper and pen.

Drafting is a messy and oftentimes exhausting stage, and you will probably go back to the planning stage to help you draft the document. During drafting, you might enlist the help of your colleagues. For example, you may have one or two colleagues read through a document before you produce the final draft of that document. Or you may have a team of colleagues collaborating with you on the drafting stages of a document.

While the drafting stage differs among individuals, there are questions that all writers can ask themselves about this stage of the writing process:

Questions writers ask themselves about *drafting*

- How will I use my outline? Will my outline be a map I use to guide me through the drafting stage? Will my outline be a springboard that I use to discover new ideas?
- Have I been given enough time to move through the drafting stage? If not, how can I proceed through this stage more efficiently?
- Can my colleagues help me pass more efficiently through the drafting stage? Whose help would be most useful to me?

One final strategy that many writers use during this stage is to allow themselves enough time for reflection. After you have been drafting, stop and think about the drafting process that you have been using. Think about what makes your drafting process successful, and why you choose to move through this drafting stage in certain ways. Also, think about what constraints or limitations you put on your own drafting process.

Reflecting during this stage of the writing process will enable you to remember what strategies were useful and what strategies may have impeded your writing success.

Chapter 11 Task Exercise

As a technical writer working within an organization, you will benefit from identifying and understanding how you proceed through drafting. To illustrate how you move through this stage, write down your responses to the following questions:

Think about a document that you have recently written (it could be a paper for school, a letter to a friend, or a report or brochure written at your workplace). Describe in writing the drafting process that you went through to create that document.

1. Are they different or similar to the phases described in this section?
2. How could you make these phases more efficient or productive?
3. During the drafting process, did you enlist the help of a colleague or friend?

Understanding how writing within an organization is unique and knowing the specific ways that you typically proceed through planning and drafting will make you a more efficient, knowledgeable, and successful technical writer.

Chapter 12

Writing Effectively by Understanding Revising, Editing, and Document Cycling

To become a good technical writer, you must understand why revising, editing, and document cycling are all important to creating an effective document.

Once you have begun to plan and draft your document, you may immediately begin one of the final writing processes—revision (Figure 12.1). Many technical writers do not acknowledge a difference between the drafting and revising stages of the writing process. For most writers, drafting and revising are performed together, or the document drafting process is followed closely by revising, which in turn influences the way that you draft.

Figure 12.1 Four stages of the writing process

Planning ⟷ Drafting ⟷	**Revising ⟷**	**Editing**
	• Revise for accurate and complete content • Revise for audience • Revise for clear and logical organization • Revise for proper mechanics, prose style	• Edit for accurate and complete content • Edit for logical organization • Edit for accurate and consistent document design • Edit for mechanical, prose style, and voice errors

In this chapter, we identify and discuss the final two stages of the writing process: revising and editing. We also discuss a type of editing strategy that is becoming more popular in the workplace: document cycling. To be a successful and productive technical writer, you must understand the answers to the following questions about these final components of the writing process:

- What do I need to know about *revising?*
- What do I need to know about *editing?*
- What do I need to know about *document cycling?*

Answering questions like these not only allows you to identify typical ways in which technical writers perform these strategies, but also enables you to start to identify the ways in which you revise, edit, and participate in the document cycling process.

What do I need to know about revising?

Revision is a process that every writer goes through, a process in which the writer rereads a document and makes changes related to accuracy, audience, organization, and mechanics. Revision is an integral part of the writing process. In this section, you will learn how to make this third stage of the writing process more successful and productive.

While we identify revision as the third of four stages of the writing process, revision is an operation that you can perform throughout the writing process. Notice that in Figure 12.1 the arrows between each stage of the writing process point backward *and* forward. The directions of the arrows suggest that the writing process is not a linear, but a cyclical, process. Therefore, you may revise during the planning stage, plan during the drafting stage, or edit during the drafting stage.

There are specific strategies you can employ to make revising more efficient and useful. During revision, successful technical writers remember to check for these four categories:

- Accurate and complete content
- Attention to audience
- Clear and logical organization
- Attention to mechanics

Accurate and complete content

As you write, one area of concern in your document is content. Revising for content is ongoing, and it is something that you can attend to throughout the writing process. Many times, writers lose sight of accuracy and completeness of the message, facts, and ideas in their document. You can ask yourself several questions about your document's content.

Questions writers ask themselves about *content*

- Is the message and point of my document accurate? Is my document's message persuasive and accurate?
- Are the facts that I am presenting throughout the document logical and accurate? Will my audience interpret the facts in the same way that I am presenting them?
- Have I explained each major point completely? Will my audience fully understand my message? How can I make my message clearer?
- What other points should be added to my message? What other discussions, explanations, or descriptions will make my message more complete?

As you continue to ask yourself these questions about content, the message and ideas that you communicate become more clear. Also, if you check the accuracy of your facts throughout your writing process, you will save yourself time during the final stages of revision.

Attention to audience

Another area of concern that you can revise for as you write is attention to audience. Revising for audience as you write enables you to continually shape the focus and content of your message to suit your audience's needs.

Revising for audience and revising for content are two strategies that are most productive if they occur in tandem. That is, while you revise for content, also make sure that the content you are presenting is suitable for your audience. For example, based on what you know about your document's audience, make sure that you provide them with just enough, rather than too much or too little, information.

The following list of questions is important to follow as you revise for audience:

Questions writers ask themselves about *audience*

- Does the document provide my audience with too much or too little information? In what ways can I reshape the document's message to make this presentation of information more suitable?
- Are the examples in my document appropriate for my audience? Will these examples need to be more clearly presented?
- Is the content of my message persuasive enough? What other strategies will my audience accept in terms of my message?
- Is my tone appropriate for my audience? Is the organization and document design appropriate as well? What changes could be made to make these more suitable?

Keeping these questions in mind while you revise for audience is a strategy that will enable you to continually improve your document. Depending upon the document you write and the communication context in which that document is situated, you may be able to ask more specific questions about audience. These questions, then, will enable you to revise even more successfully for issues relating to audience.

Clear and logical organization

Another area that should be attended to throughout the writing process is revising for clear and logical document organization. As a technical writer, you probably create a detailed outline of your document's major points during the planning stage. Of course, this outline helps you to make organization decisions throughout the writing process. However, as you continually revise for content and audience issues, your document's organization may need to be revised as well.

One of the best strategies for revising for organization is to clearly identify your document's objectives and major points. Understanding these points from the beginning of the writing process makes revising for organization more efficient and productive.

To revise most effectively for the clarity and logic of your document's organization, ask yourself the following questions:

Questions writers ask themselves about *organization*

- What is my document's objective and major points? How well does my organization frame and support these points?
- What other strategy could be used to better organize the document or a section of the document?
- Will this organization be clear to my audience? Will my audience find this organization persuasive?
- Is this organization logical? Are there any points that are not logical or may be interpreted falsely?
- Is there a more efficient means of organization? Will my audience become bored or confused by this organization?

Technology, such as word processing software, simplifies revising for organization. Writers can easily rearrange and reorder text in a document by blocking, cutting, and pasting information from one section of a document to another. Also, some software applications allow you to move back and forth with ease from the draft of the document to its outline. This process allows you to refer easily to your organizational outline for ways to reshape and revise your document.

Attention to mechanics

A final strategy that you can use to revise your document is to attend to mechanics throughout the writing process. Mechanics include a variety of sentence-level concerns such as spelling, grammar, usage, word choice, and style. Chapter 18 provides you with good starting points for revising for these elements.

Frequently, writers believe that they do not need to attend to mechanics *throughout* the writing process, thinking that it is more efficient to correct these errors during editing. However, writers can edit much more easily and productively by correcting mechanical errors throughout the writing process.

Response Exercise 12.1

As a technical writer working within an organization, you benefit from understanding how you revise. To illustrate how you move through this stage of the writing process, write down your responses to the following questions:

1. What four categories should you check for during revision? Identify two important questions that you should ask yourself about each category.

2. Why is it important for you to revise for all four of these categories? Would document quality suffer if you did not revise for all four?
3. Think about a document that you have recently written (it could be a paper for school, a letter to a friend, or a report or brochure written at your workplace). Describe in writing the revising process that you went through. Is this process different or similar to that process described in this section? How could you make this process more efficient or productive?

What do I need to know about editing?

Like planning, drafting, or revising, editing is an important stage in the writing process. In this section, you learn many of the basics of editing, including strategies that successful editors use. Remember, the strategies discussed here are useful general guidelines rather than hard-and-fast rules.

Editing, an integral part of the writing process, occurs as one of the final steps of the writing process. Many writers do not edit their own documents; rather, editors often step in during this stage of the writing process.

Whether or not you perform your own editing, understanding what makes the editing process successful is important for any technical writer. Many of the elements that editing attempts to correct are similar to the kinds of elements that writers revise for throughout the writing process. Therefore, the more efficiently and effectively one has revised the document, the easier that document is to edit.

Successful technical writers ask several questions about editing a document. Note that the strategies suggested by these questions are general enough to be used for nearly every type of document:

Questions writers ask themselves about *editing* a document

- Are the message and content of the document accurate? Are the message and content organized logically?
- Is the document design accurate? Is the document design consistent?
- Do the mechanical errors detract from the professional tone? Do these errors detract from the message and content?

Editing is frequently a process that is carried out by someone other than the document's primary writer or writers. Many organizations, however, make the editing stage highly collaborative, and editing becomes the responsibility of a number of different people within the organization. Whenever a group or a network of colleagues contribute their suggestions to editing a document, this process is called *document cycling*.

Response Exercise 12.2

As a technical writer working within an organization, you will benefit from understanding how you edit the documents that you produce. To illustrate how you proceed through the editing stage, write down your responses to the following questions:

1. What four categories should you check for during editing? Identify two important questions that you should ask yourself about each category.
2. Think about a document that you have recently written (it could be a paper for school, a letter to a friend, or a report or brochure written at your workplace). Describe in writing the editing process that you went through. Is this process different or similar to that process described in this section? How could you make this process more efficient or productive?

What do I need to know about document cycling?

At many organizations, document cycling is an important part of writing process, and it is a collaborative method for performing editing. Document cycling is a way for a writer to receive suggestions from a variety of readers about how to improve the document before that document is submitted to its primary audience. When a document in its final stages of the writing process is sent to multiple readers for suggestions, that document is part of document cycling.

Document cycling can involve a variety of colleagues from different departments within the organization. Colleagues who contribute to the document cycling process may include editors or other technical writers; personnel with specialized technical knowledge, such as engineers or designers; personnel who work within the budget or finance departments; lawyers or other legal personnel; or supervisors who have business or managerial degrees.

Including a broad range of expertise in this process produces a rich and diverse set of commentary and suggestions for document improvement and change. However, having a diverse group of reviewers to suggest document revisions can lead to inefficiencies, because the process is time consuming and a writer often receives contradictory recommendations about changes that reviewers want. The organization that promotes document cycling as a viable part of the writing process must also have an efficient method for transferring the document from the writer to the reviewers. One of the best ways to increase efficiency within document cycling is to pass along the document from writer to reviewers via e-mail.

Writers may receive feedback from colleagues in a number of different ways and must be alert to the best ways to receive the feedback. For example, you may receive suggestions written on the document itself or on a separate memo or note attached to the document. Also, you may receive feedback over the phone in an informal conversation or in person during a meeting. Regardless of the method of feedback, you must be ready to take careful notes about the revision suggestions that your colleagues provide.

Chapter 12 Task Exercise

As a technical writer working within an organization, you will benefit from identifying and understanding the importance of document cycling. To illustrate why this process is important, write down your responses to the following questions:

1. What is document cycling? How is it used in organizations?
2. How is document cycling different than revision?
3. Typically, what kinds of colleagues are involved in the document cycling process?
4. What are three important advantages of document cycling?

Whether you receive editing suggestions from one colleague or participate in document cycling, you make the ultimate decision about what suggestions to use in your document. Understanding how to revise and edit are just as important as understanding how to plan or draft. Taken together, planning, drafting, revising, and editing are four important stages in the writing process that all contribute—when performed effectively—to the success of the document.

Chapter 13

Testing Documents for Accuracy and Usability

To become a good technical writer, you must understand that many documents benefit from usability testing. Understanding the importance of usability testing, the common methods of testing, and the ways to make testing in the workplace as successful as possible helps you to create useful and accurate documents.

As a technical writer, you want to create documents that are accurate, usable, and clearly written. One of the best ways to create documents that exhibit these characteristics is to perform usability testing. Usability testing is a way for document writers—and writers of other resources, such as Web sites—to watch, listen, and understand how readers use the products that they have created and to then make changes to that product before it reaches its actual audience.

Usability testing, when implemented correctly, can be one of the most successful and profitable programs at any organization. As a technical writer, you will be involved in such a program and may even be the instigator of that program. Therefore, you need to know the answers to the following questions about usability testing:

- Why is usability testing *valuable*?
- What are the common *types* of usability testing?
- How do I make the *usability testing procedure successful*?

Understanding how your documents can benefit from usability testing is valuable knowledge for any technical writer to possess. You must clearly understand the benefits of a usability testing program, not only for your own benefit, but also to persuade your colleagues to support and fund a usability testing program at your own organization.

Why is usability testing valuable?

A document that is usable is one that readers can read and use quickly and efficiently. Many times, if a document, such as an instruction manual for a lawn mower, is highly usable, it will take the reader of the manual (in this case, the user of the lawn mower) less time to understand how to operate and maintain that equipment. If a Web site describing a vacation locale is highly usable—that is, if it provides the necessary information, is well organized, and anticipates reader use—then that site may persuade readers to vacation there.

Usability testing is a process that helps the writer of a document or other resource assess the usability and accuracy of that document before it reaches its primary audience. Based on the results of the test, writers assess the document's usability and accuracy and revise it accordingly.

Technical writers use three different types of usability testing (we discuss all three types in length in the next section): text-based, expert-based, and user-based. Understanding which type(s) of testing are most useful and appropriate for your document—and knowing when to perform what type(s) of testing—makes your document more usable, accurate, and successful.

Usability testing is most commonly performed during the instruction manual writing process. In this case, usability testing allows the writer to revise a product's instruction manual based on watching, listening, and interpreting how a user reads and follows the instructions. Similarly, the results of usability testing guide the writer's revision of that manual for technical accuracy, accessibility, and clarity.

Usability testing is equally valuable for a host of other types of documents. For example, documents and resources, such as product/service information brochures, Web sites, proposals, reports, and employee handbooks, can all benefit from usability testing.

Usability testing, whether it involves an instruction manual or a Web site, should be an integral part of all stages of the writing process. It should also be

> **Samuel's story…**
>
> As a technical writer for a small Web site design company, Web-Serve, Samuel is assigned to create an on-line tutorial for a software program. This program, created by a successful computer software company, Library.com, enables its users to organize, list, and perform searches for a large body of reference texts, such as articles and books. The program, while very useful for students and writers, was difficult to master for first-time users. Therefore, the on-line tutorial that Samuel creates must consider the variety of needs of these novice users.

conducted throughout the writing process, from drafting through revising. Figure 13.1 shows the four stages of the writing process. Depending on the type of document you are writing, you may choose to include usability testing during the *drafting* stage and again during the *editing* stage.

Figure 13.1 Four stages of the writing process

Planning ↔ Drafting ↔ Revising ↔ Editing

Drafting:

- May use outline or notes to help generate text
- Continually make communication context, content, organization, prose style, and design decisions
- May collaborate with others to help generate text

Editing:

- Edit for accurate and complete content
- Edit for logical organization
- Edit for accurate and consistent document design
- Edit for mechanical, prose style, and voice errors

In their book presenting practical techniques and useful approaches to usability testing in any organization, Dumas and Redish (1993) identify several benefits that companies gain by including a usability testing program (14). According to Dumas and Redish, virtually any type of organization can see positive results in these three areas:

- Company reputation
- Product costs
- Personnel costs

Company reputation

Company reputation is positively impacted by a rigorous usability testing program. For example, usability testing an on-line tutorial for a software program helps the writer to produce an accurate and usable tutorial, which allows software users to easily and quickly use the tutorial and the software itself. Satisfied users tend to spread their product satisfaction by word of mouth, thus strengthening your company's reputation.

Product costs

Product costs can be lessened by a rigorous usability testing program. For example, usability testing can be used to test a new platform for your company's electronic database before that database is installed on all of your company's computers. In this case, a rigorous program of usability testing not only increases employee satisfaction with the usability and accuracy of the platform, but also decreases overall product costs, because the platform will be designed correctly before it is installed.

Samuel understands the importance of the tutorial...

Samuel understands that producing a usable and accurate tutorial will benefit both his Web site design company, Web-Serve, and the company that has designed the computer program itself, Library.com. If Samuel's tutorial is successful, Library.com's customer satisfaction—and the company's reputation—will grow. Because of this success, Library.com will probably continue doing business with Web-Serve.

Usability testing decreases other product costs, such as software updates or maintenance releases. When your employees or the customers who have purchased your product purchase one that is "buggy," or has errors, they need to be sent an update. Sending updates to employees or customers is a product cost that your company should not incur.

Personnel costs

Personnel costs, such as the costs of maintaining a customer service line or training new employees to use a new product or service, can be lessened by a rigorous usability testing program. For example, usability testing can be used to test an instruction manual for the lawn care product your company manufactures before that product is sold. In this case, a rigorous program of usability testing not only will increase customer satisfaction, but also will decrease overall personnel costs because the customer service program may become less necessary.

Usability testing not only improves the documents you create, but also benefits your company's reputation and decreases product and personnel costs your organization may incur. As a technical writer, you may be in the position to persuade your organization about the benefits of such a testing program, or you may be involved in organizing and maintaining the program. Regardless of your role, it is important that you understand the different types of usability testing programs.

Response Exercise 13.1

Identifying possible audiences for the documents you create is important to planning that document and performing usability testing. Knowing all of your document's possible audience types enables you to better plan that document's usability testing.

For each of the *subjects* below, list the different audience types who may be interested in that subject. Then write down your responses to the *questions* about these audience types.

Example: Possible audiences interested in the subject, "changes in bicycle helmet safety standards," include young adult/adult bicycle riders, parents, bicycle helmet manufacturers, and bicycle store owners.

Subjects

a. Environmental impact of the new paper mill
b. Violence ratings for video games
c. Financial support for local symphony
d. Mistreatment of the elderly at an area nursing home

Questions

1. Describe in a sentence or two what specific interest each audience type has in the topic.
2. Choose two of the audience types that you have listed. How is each type's interest different? How is each type's interest similar?

What are the common types of usability testing?

As a technical writer, you may be involved in many aspects of the usability testing that occurs at your organization. Therefore, you need to understand the different usability testing options. The three major types of testing that Schriver (1988) categorizes are discussed in this chapter:

- Text-based
- Expert-based
- User-based

In this section, you will learn the definition, characteristics, and purposes of each type of usability testing. After reading this section, you should not only understand the differences of each type of testing, but also be able to explain those differences and purposes to others.

Text-based

Text-based usability testing involves assessing the textual and visual features of a document—literally the words, sentences, and visual aids that make up the document. This type of testing is most useful for evaluating how effectively and efficiently an audience can read the text.

Text-based testing is performed by a person (the writer, an editor, a colleague) or by a computer. Many software programs now offer a spelling and grammar check option. These features qualify as text-based testing methods because they assess word and sentence structure quality.

Samuel decides that usability testing must be performed…

Before Samuel begins drafting the on-line tutorial in earnest, he makes a decision about usability testing. Samuel decides that usability testing must be an integral component of the on-line tutorial writing process. This is the first large tutorial project that Samuel has ever been assigned, and he has not performed usability testing on a document before. Samuel accesses and reads two texts about usability testing—one book and one article—that Web-Serve has available in its company library.

Other forms of text-based testing include assessing documents using field-specific or company-mandated style guides. Field-specific guides may include the *Chicago Manual of Style;* this guide provides writers with rules for word usage, sentence structure, abbreviations, and many other elements. Your organization may adhere to its own style guide, either a guide that is widely published like the guide mentioned above or a company-published style guide.

Text-based testing can be used for every type of document you produce. This type of testing is useful for assessing how easily and efficiently a document can be read. However, text-based testing takes the words and sentences out of context. Therefore, text-based testing should not be used exclusively to assess a document's accuracy or usability. Using text-based testing in conjunction with another testing measure, like one or more of those identified below, produces thorough, accurate results.

Expert-based

A second type of usability testing, expert-based, may involve one or more levels of review—from formal to informal readers. On one level, *technical readers* assess the document for technical and content accuracy. On a second level, *style readers* assess a document's overall communicative accuracy. Style readers, usually experts within a writer's field, read a document for broader issues, such as completeness, conciseness, clarity, and attention to audience.

One document may pass through these two levels of expert-based review. For example, an employee handbook written for an audience of part-time outdoor concert workers contains a section about ensuring patron safety. Specifically, this section shows workers how to recognize and combat sunstroke in concertgoers. *Technical readers* of this section of the handbook assess the document for medical accuracy. For example, sunstroke symptoms must be properly identified and described, and the necessary treatment methods must be included in the proper order. *Style readers* assess this section of the handbook for ease of readability, consistency, and clarity. For example, visual aids must be provided when necessary, and the language describing the symptoms must be consistent and clear.

Another form of expert-based review that you may encounter at your workplace involves the legal reader. The legal reader assesses whether or not the information you have provided your readers contains any negative safety or health implications.

Expert-based testing may be used for nearly every type of document. This type of testing is useful for assessing a document's technical accuracy, its stylistic and audience readability, and its legal soundness. Expert-based testing, in one or all of its forms, is one of the most popular types of usability testing in the workplace.

User-based

The final type, user-based usability testing, may be conducted in two different ways. Either you watch and listen while someone reads aloud and uses the document, or you ask the user questions about the document's usability after he or she has read and used the document. Most frequently, user-based testing is used to assess the accuracy and usability of instruction manuals.

For example, an instruction manual written to help users assemble a child's bicycle benefits a great deal from user-based testing. In this test, someone who has never read through the instructions before reads the instructions and actually assembles the bicycle. The user reads the manual out loud and talks through the assembly steps as they are performed. The user is asked to comment on the manual and to the activity of assembling the bicycle during the process itself.

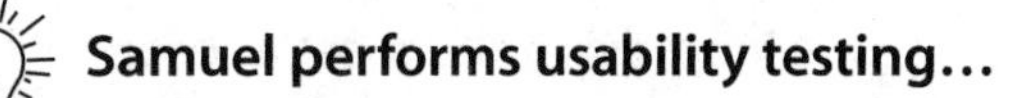

Samuel performs usability testing…

After reading the materials on usability testing, Samuel asks a colleague who has performed user-based testing in the past to help him with the on-line tutorial testing. Samuel decided to perform user-based testing to insure that the process is clear for future users of the tutorial.

There are several questions writers ask themselves while conducting user-based testing:

Questions writers ask themselves about *user-based testing*

- Have I recorded the user on tape so I can go back and listen to and analyze the comments the user makes?
- Have I made the user comfortable? Have I explained the user-based testing procedure adequately?

- Do I have the document in front of me so I can make notes on it while the user reads?
- Have I prompted the user to read and talk out loud?
- Have I observed enough document users? Am I able to identify patterns among all of the users?

The second method of user-based testing has the user read the document and perform the task without reading or talking aloud. When the task is complete, the user is asked questions about the document and the experience. These questions may be in an interview format or in the form of a survey.

Both methods of user-based testing assess the accuracy and usability of many kinds of documents, including instruction manuals. Having people who have never seen the document before attempt to follow its instructions is one of the most telling tests a document can be put through.

Samuel performs usability testing...

Samuel and his colleague oversee the user-based testing of the tutorial over a period of three afternoons. During this time, Samuel watches, listens to, and analyzes ten users of his on-line tutorial program. Samuel uses a check list and a testing script to remind himself to set up the testing area correctly for each user and to explain the testing procedure similarly to each participant.

Response Exercise 13.2

Identifying the most appropriate and useful usability tests to perform on a given document is the first step toward creating a successful usability testing program.

For each of the *documents* below, list the kind(s) of usability tests—text, expert, or user based—that you believe will be most useful to perform. Then for each document, respond to the *questions* that follow.

Example: For an "instruction manual for a hand-held grass trimmer," text- and user-based testing could be performed.

Documents

a. On-line tutorial for a computer software program
b. Instruction manual to accompany a combination coffee bean grinder/coffee maker
c. Informational brochure describing the policies and procedures of a new home-to-clinic shuttle service available for patients with limited transportation options
d. A Web site that describes and promotes a county-run recreation vehicle campground

Questions

1. Describe, in one or two sentences, the kind(s) of testing that you have selected. Do these tests involve just the document itself, an expert, or a user?
2. What are the benefits of the kind(s) of testing that you have selected? How would you persuade your supervisor to fund testing like this?

How do I make the usability testing procedure successful?

It is important to understand why usability testing is often necessary and what types of testing exist. However, technical writers often have control over the testing procedure itself. Four strategies can be used by technical writers to ensure that the usability testing procedures at their organizations are as successful as possible. The four strategies that you will learn about in this section are practical and relatively straightforward and can be used at any type of organization:

- Decide on appropriate usability testing methods
- Build testing into the writing process
- Communicate the testing results to the writer
- Lobby for usability testing support

Decide on appropriate testing method(s)

One important strategy for making usability testing successful is to decide what testing method(s) you will use. Will you perform text-, expert-, or user-based testing? If you decide on user-based testing, ask yourself: "Will I use a read- and talk-aloud strategy?"

Another important component of this decision making process is assessing the availability of testing resources. Several elements impact the testing process: budget, time, personnel (to conduct testing), and users (to participate in user-based testing procedures, for example). Deciding what testing methods are appropriate for different project situations is important, and there are several questions that writers ask themselves about these types of testing decisions:

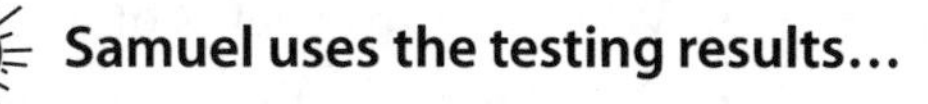

Samuel uses the testing results...

The results of the user-based testing of the on-line tutorial were successful. In fact, Samuel believes that he could not have written the tutorial without this type of testing procedure in place. Samuel discerned a pattern in the responses his participants gave. Samuel used these responses to make the final, important revisions to his on-line tutorial.

Questions writers ask themselves about
appropriate testing method(s)

- What eventual outcome do we need from the testing? Do we need to assess document style? Technical accuracy? Do we need a thorough assessment of usability?
- In the past, have we conducted testing on a document such as this? If so, was the testing successful? What did we learn from the testing?
- Will one type of testing be adequate to assess what we need? Do we need to include more than one type for this document? If so, what types?
- What financial resources do we have to conduct testing? Do we have enough time? Do we have enough personnel and users (if necessary) to conduct the testing?

Build testing into the writing process

A second important strategy to keep in mind when you implement usability testing is to build testing into the writing process (Figure 13.1). Once you have begun the planning stage of the document, you must also begin planning for usability testing.

To plan for testing, you must decide (a) if testing is necessary for this document and, if so, (b) what testing methods are appropriate for the document. Also, you must keep in mind your answers to these important questions:

Questions writers ask themselves about
testing and the writing process

- Will testing be a part of just one stage of the writing process—such as the editing stage—or will it occur more than once in this process (perhaps during drafting and editing)?
- What are the benefits of including testing more than once?
- Can this project support the financial resources, time requirements, and personnel needs of testing that occurs more than once during the writing process?
- How can I persuade my project team members/co-workers/supervisors that testing is important and should be built into the writing process?

Communicate results effectively

A third important strategy for making usability testing successful occurs after the testing has been conducted: communicating the results to those who need them.

Usability testing is only effective if its results are implemented and the document is improved.

An organization that has recently begun usability testing or is conducting it for the first time also needs to make sure that an effective communication network is in place. Establishing a communication network that connects the personnel who conduct the testing with those who must use the results is an important aspect of testing.

Lobby for testing support

Usability testing is time consuming and can be expensive. However, its long-term benefit is not simply document success, but company success. Implementing testing at your organization may be difficult, but as a technical writer, you are well aware of the benefits of usability testing, and it is your job to lobby for testing support and show your co-workers the value of usability testing.

Actually conducting a usability test can be a good way to pique your co-workers' interest in usability testing. Have co-workers—whether or not their product is being tested—observe one of the testing procedures, such as user-based testing. Observing how and why a document fails or succeeds can be an eye-opening and interest-generating experience. There are other ways to lobby for testing support, and writers often ask themselves these questions about this topic:

Samuel lobbies for increased usability testing...

Samuel feels very confident in the way he included usability testing in the on-line tutorial writing process. While the success of the tutorial with actual customers remains to be seen, Samuel lobbies for increased usability testing at Web-Serve. Samuel discusses his own positive experiences with his fellow technical writers, his supervisor, and his manager. He argues that increasing testing funding will lead to greater Web-Serve customer satisfaction.

Questions writers ask themselves about *lobbying for testing support*

- Do your co-workers know the value of document testing? Why is co-worker support valuable?
- What methods—including having co-workers observe testing—can you implement to garner support for testing?

Chapter 13 Task Exercise

As a technical writer, you must not only understand the uses and benefits of usability testing, but also how to communicate these elements to others. Read through the scenario and write down your responses to the questions that follow.

Scenario

You work as a technical writer at a company, Design-Ware, that designs and sells computer software. Your company is small and relatively new and has never had a usability program in place. Recently, Design-Ware has been receiving a number of complaints from customers about a newly released software program. These complaints involve customer confusion about program operation and particularly about using the on-line instruction manual. You believe that if the program and its manual had included usability testing, this problem would have been caught before the program was released.

You know that your supervisors do not know about usability testing but that now is the time to inform them. At the next staff meeting, you plan to give an oral presentation introducing the staff and supervisors to the benefits of usability testing. To prepare for the presentation, you have asked a fellow technical writer to play "devil's advocate," and to devise a list of questions that your co-workers may ask you during the presentation about usability testing. As a believer in the benefits of usability testing, you must prepare complete and persuasive responses to the following questions prepared by your co-worker:

1. Identify the different kinds of usability testing that we would ideally perform on an on-line instruction manual for a new software program.

   Describe these types of testing to me in language that I, a business manager, can understand.
2. Writing and revising an on-line instruction manual like this one is time intensive and costly and usability testing just adds to these factors.

   Persuade me to include usability testing in this writing process—for example, what are the benefits that the product, customer, and our company will receive?
3. As a technical writer, you would be in charge of conducting user based testing for an on-line instruction manual, and you have stated that this testing would take up several afternoons of your time (more or less based on the number of participants).

   Persuade me to assign you exclusively to this project by explaining the steps of a typical user based testing procedure.

Chapter 14

Working With a Designer and Printer to Produce Documents

To become a good technical writer, you must learn to communicate with your colleagues. Two colleagues with whom you may collaborate a great deal, the designer and the printer, influence the appearance, appeal, and usability of your document.

Depending on the type of document you create, you may need to collaborate with a publication designer, a printer, or both in order to produce a quality document. A publication designer works with the editor and the writer to "coordinate art and typography with content" in order to make documents as appealing and usable as possible (Nelson, xiii). A printer may work with the editor, writer, and designer—or just with the writer—on the production phase (the printing and distribution) of the document.

Lengthy documents or documents with a complicated design or layout benefit from the services of a designer, while documents with a large distribution, or in which desktop publishing is not feasible, demand the services of a printer. Depending on the types of documents you write, you may work closely with a single designer and a single printer. Therefore, you may have the opportunity to develop a close working relationship with one or both of these colleagues.

If you work for a relatively large organization, you may have a designer or a printer working at your company. However, both the designer and printer are usually consultants who work for independent agencies, and they are not on staff at your organization. Regardless of their location, developing a good rapport with your designer and printer will greatly assist your ability to produce professional looking and aesthetically pleasing documents.

To accomplish this relationship, you must have a good understanding of the work performed by these professionals. Consider the following questions about their work:

- What roles do design and printing play in the *writing process*?
- What should I know about the role of the *designer*?
- What should I know about the role of the *printer*?

As a writer, knowing basic information about both design and printing will serve you well. Whether you communicate with a designer and printer frequently, or only once or twice a year, this knowledge helps you to communicate more effectively with them, and in the process, enables you to produce more appealing and usable documents.

What roles do design and printing play in the writing process?

You have learned about ways to plan, draft, user test, revise, and edit the documents that are typically produced by technical writers. While the importance of collaboration is a common aspect of this discussion, you have not yet learned about how and with whom writers collaborate after the editing stage of the writing process (Figure 14.1).

Technical writers collaborate with their colleagues in a number of ways. For example, members of a project team producing a recommendation report may collaborate on all stages of the writing process—from planning through editing. In another example, as the primary writer, you may ask three colleagues, during the document's editing stage, to read and provide feedback on the draft of a brochure. However, after the planning, drafting, revising, and editing stages of the writing process, your document may pass through two more stages, as shown in Figure 14.1.

Figure 14.1 The design and production stages

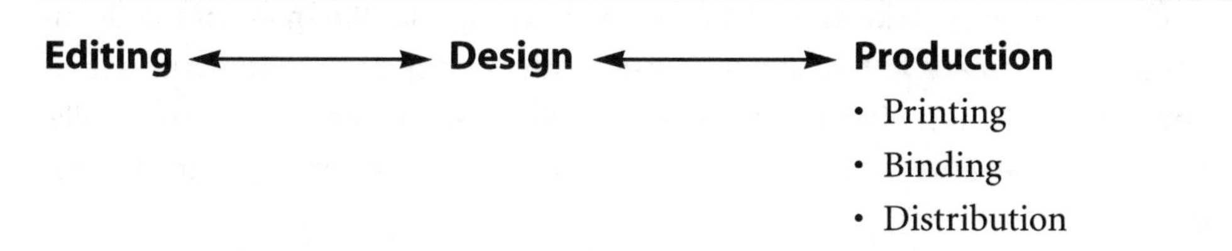

These final two stages, design and production, may not occur with every document that you create, but when it is necessary to move through these stages, you turn to a designer and a printer, respectively. Interestingly enough, with the popularization of desktop publishing, the line between the design and production stages—particularly the document design and printing processes—has become less distinct.

Rather than using the services of a designer and a printer, large and small companies in many different sectors—from manufacturing, business, and finance to computers and non-profit—may use desktop publishing as an effective and cost-efficient means to produce high-quality documents. Any organization (or any individual, for that matter) may assemble a desktop publishing unit by purchasing a computer, page composition software, a scanner, and a laser printer.

To produce quality documents using a desktop publishing unit, the writer works at the *computer* with *page production software*, such as Aldus Corporation's PageMaker, which designs the look of the page by enabling users to incorporate professional page designs into nearly any type of document. A *scanner* scans and inputs photographs or other illustrations into the document, while a *laser printer* produces documents that have a high-quality, professional appearance.

By enabling the writer to perform the work of professional designers and printers, a desktop publishing unit may suffice for producing most documents. However, there is still a place for the designer and printer in the production of many types of documents. In the next section, we examine the role of the designer and outline five basic design principles that will help you, as a technical writer, communicate more effectively with your designer.

Allison's story...

As a technical writer for SunSpot, a large agricultural equipment manufacturing corporation, one of Allison Holcomb's responsibilities is to edit the company newsletter. This newsletter, published monthly, has a circulation of roughly 2,000. The newsletter's audience consists of current and retired SunSpot employees from all sectors of the corporation—from equipment design and manufacture to sales and service. For the newsletter, Allison solicits and receives articles not only from SunSpot employees, but also from writers and those interested in agriculture who are not employed by SunSpot. Allison has been asked to change "the look" of the newsletter to reflect the recent SunSpot theme, "Equipment for farming in the new millennium."

Response Exercise 14.1

For technical writers, desktop publishing is a key component in many of the documents that they create. To familiarize yourself with desktop publishing, write a family newsletter that describes the current activities of each member of your family, and the special events that your family has participated in throughout the past year.

To create your family newsletter, access and use desktop publishing equipment identified above. Using the page publication software's tutorial and help manual, experiment with as many of the format and design features as possible. Once you have completed your newsletter, write down your responses to the following questions:

1. What design elements does your newsletter contain?
2. What page publication software features were most useful when creating your newsletter?
3. Identify three design elements of your newsletter that could only be created on the page publication software. Do these elements make your newsletter appear more professional?

What should I know about the role of the designer?

While technical writers learn the basics of visual communication and the elements of page and document design, they often call upon the expertise of a professional publication designer to create a high-quality document. A designer is a valuable resource because he or she not only understands the importance of the document's communication context and message, but also is able to translate these aspects of the document into a specific set of sophisticated design features. With any number of important documents, a designer's training and experience helps you to create a very attractive and functional document.

As a technical writer, you may know what design features to include in the document, but many times the designer knows the best ways to incorporate those elements into the document, or may know of even more effective design ideas. In his book, *Publication Design*, Roy Nelson describes five principles of design that,

Allison decides to consult with a production designer...

The first changes Allison makes to the SunSpot newsletter are to its design. After the new design is decided, Allison will gradually revise the focus of the articles in the newsletter to better reflect the new theme. To do this, Allison will begin by using several selected articles that focus on farming for the future. For now, though, Allison must concentrate on revising the current newsletter design. Through a colleague's recommendation, Allison decides to consult Kerry, a production designer, to get ideas for the newsletter's new "look."

"consciously or unconsciously, designers tend to operate by," and that "apply to all forms of art," not just documents (30):

- Balance
- Proportion
- Sequence
- Unity
- Contrast

The placement and use of these principles within a document depends largely on the document's topic and communication context. Therefore, understanding each principle's definition, and knowing how each impacts document design, allows you to communicate with the designer to create the best possible document.

Balance

The design principle of *balance* concerns itself with the weight of the text and visual aids on the page, ensuring that everything placed on one side of the page balances with what is on the other side of the page. Typically, designers produce two types of balance on the page: bisymmetric and asymmetric.

A *bisymmetric* balance evenly and consistently balances all of the textual and visual features of a document. For example, Figure 14.2 is a thumbnail sketch of a two-page document that exhibits bisymmetric balance. Notice how all elements—even the title (the shaded box at the top of the document)—are centered across both pages.

Figure 14.2 Thumbnail sketch of a document exhibiting bisymmetric balance

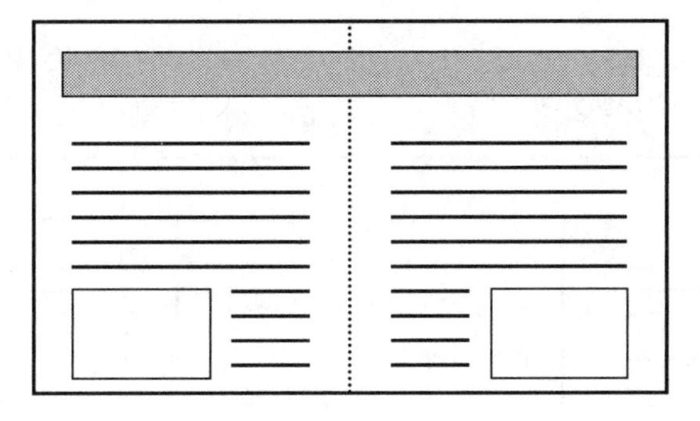

Figure 14.3 shows a two-page document that exhibits *asymmetric* balance. A document that is asymmetric is still balanced. However, its features are not evenly and consistently centered or balanced on the page. In Figure 14.3, for example, the title (the shaded boxes) is not evenly centered across both pages, yet it is placed on the page in a balanced, asymmetric fashion that is pleasing to the eye.

Figure 14.3 Thumbnail sketch of a document exhibiting asymmetric balance

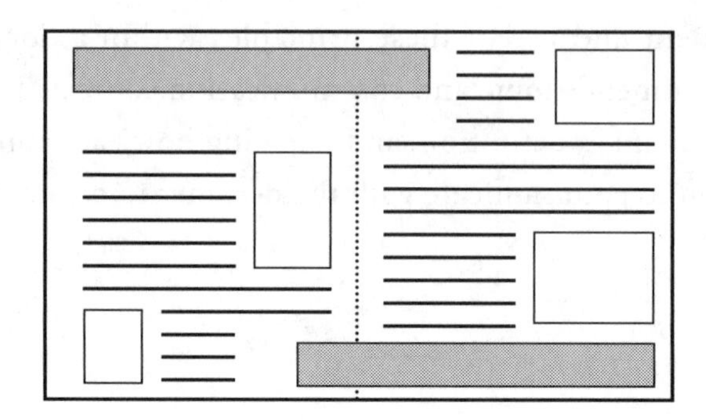

Proportion

Proportion, a principle taken directly from fine art, concerns what dimensions readers find most pleasing on the page. Readers find proportions with ratios of 2-to-3 or 3-to-5, rather than strict 1-to-1 ratios, most pleasing to view and read. Ideally, all proportions, from the white space surrounding illustrations and text, to the document's margins and the page size, should correspond to a 2-to-3 or 3-to-5 ratio.

In terms of proportion, readers find Figure 14.4a less comfortable to read than Figure 14.4b because the latter figure visually splits the page, not in half, but along a 2-to-3 ratio.

Figure 14.4a Page showing a 1-to-1 ratio

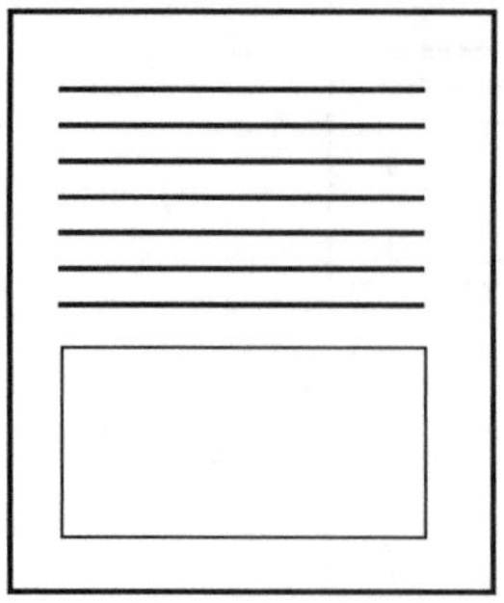

Figure 14.4b Page showing a 2-to 3-ratio

Sequence

The principle of *sequence* is defined by the fact that readers from Western cultures read text from left to right. Therefore, Western readers first look to the left side of the page and follow any text or visual aids from left to right.

Sequence is influenced by other standards that govern how readers read. For example, besides moving from left to right, a reader's eye tends to move from:

- Large to small
- Black to white
- Color to non-color
- Unusual shape to usual shape

Sequence, then, is the design principle that directs the reader's eye through the document. Successful designers know that to best use sequence in a document, the reading tendencies of the reader must be kept in mind. Documents that use sequence effectively work with, not against, the ways that readers read.

Unity

Unity is a design principle that concerns all of a document's design elements. To exhibit unity, all design elements must not only have a unifying and complementary theme, but also must support and complement the document's content and communication context.

Design elements that create unity within a document when combined effectively include typeface; arrangement of visual aids and text; use of white space; and use of lines, boxes, shading, and color. The choice and placement of each of these elements must be purposeful and consistent, while all elements must combine to create a document's specific "look and feel." For example, specific design elements may be chosen that, when combined with the document's topic and communication context, create either a conservative, modern, futuristic, or rustic look.

Another way designers achieve unity within a document is to pair a particular artist's or photographer's work with a specific document. One photographer's vision, technique, and subject matter may complement the message and communication context of a certain document. Creating unity among text and illustrations or photographs also means using one photographer or illustrator's work throughout the entire document.

Contrast

A designer uses the principle of *contrast* to create a focal point on the page. This is achieved through the use of visual aids, color, or the grouping of related visual aids and text. The principle of contrast, or successfully creating one convergence point or focus, is used to give readers a reference point in the document.

The design principle of contrast is effective only when a single contrast element is used. Creating contrast with more than one focal point confuses the reader and reduces the visual effect. For example, placing a swath of color at the lower right-hand margin and an unusually shaped, boldly drawn illustration at the opposite corner forces readers to decide where on the page to focus their attention. For the reader, this type of decision-making causes confusion. Therefore, rather than creating competing sites of contrast, create one site on which readers may focus.

Allison works with the designer…

As a designer, Kerry has worked with other writers on similar types of projects. Through her consultation with Kerry, Allison makes several important design changes to the newsletter. She and Kerry decide to use the "new millennium" theme to unify all of the design features. The new elements of the newsletter's design include typeface, the use of specific visuals, and the arrangement of visuals and text, which all suggest a futuristic "look." Kerry also suggests a new way to balance the articles with those visual aids included in each issue. Changing the elements of balance and proportion will now enable Allison to use this document's space more efficiently.

Understanding these five design principles enables you to better communicate with the designer about the look you want to create for your document. Being able to speak knowledgeably about these principles not only allows you to communicate more effectively and to develop a good rapport with the designer, but also enables you to produce better, more visually appealing documents.

Response Exercise 14.2

Understanding the designer's role in the production of a document means knowing the five principles that designers follow. To help you learn more about these design principles, collect four or five magazines of different types (home interior, sports, fashion, news, art).

While paging through each magazine, closely examine the types of design features that are present in two types of article: an opinion article or column that appears regularly and that magazine's feature article. List the design elements that make up each article. For example, list the approximate margin widths, the placement and arrangement of the text and visual aids, the use of color, the style of typeface, the format of the headings.

Write down your responses to these questions about each article's design features and what those features say about the magazines you have selected:

1. Does each article exhibit asymmetric or bysymmetric *balance?* Create a thumbnail sketch of each article to help you decide.

2. Does each article effectively exhibit the principle of *proportion?* What elements in the article—textual, visual, or both—use a 2-to-3 or a 3-to-5 ratio, rather than a 1-to-1 ratio?
3. As a reader, what features in each article do you notice first? List, in *sequence*, the features that you notice in each article.
4. Does each article exhibit *unity?* If so, list the design elements that help to unify each article.
5. What feature is the focal point of each article? Is this use of *contrast* effective? Why or why not?
6. After having read and responded to the questions for two or more magazines, describe why different types of magazines might have different design features. For example, what design features distinguish a home interiors magazine from a sports magazine? What do these features say about the message or vision of that magazine?

What should I know about the role of the printer?

If a technical writer does not have access to a desktop publishing unit, or if the document is too large or complicated for desktop publishing to produce in an effective and cost efficient manner, the services of a professional printer are needed to complete the job. Developing a good rapport with your printer, whether you use his or her printing services once each year or once each month, is important for you professionally, and for document quality.

As a writer, you know what the completed document should look like, but communicating this to the printer is often difficult, especially if you do not know what work the printer performs or what special vocabulary the printer uses. Typically, printers are involved in three aspects of the production process: printing, binding, and distribution.

As you become more comfortable working with your printer, or if you send the same type of document to be printed on a regular basis, you may be able to forego formal discussions about your production process needs. In these cases, you may or may not choose to include a note paper-clipped to the copy specifying the printing, binding,

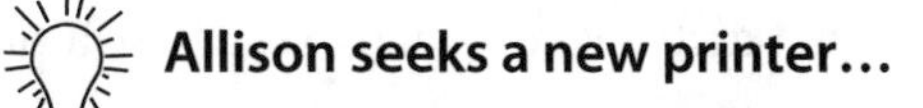

Allison seeks a new printer...

Allison is not sure that SunSpot's current printer can handle the print job of the updated newsletter. During the past months, Allison found that the representative at the printer who handles SunSpot's account has not been very responsive, nor has she treated their business with the respect that a returning, long-term client deserves. Allison decides to look for a new printer. To begin with, Allison provides several prospective printers with the brochure job specifications and requests a price quote from each. Allison also decides to meet with each potential printer in person to assess how comfortable she feels with each representative and his or her business environment. As a representative for SunSpot, Allison makes a decision based on price, turnaround time, service, and comfort.

and distribution requirements. More often than not, your printer has all of the production specifications for your document(s) on file, and no communication is necessary.

However, if you have not worked with your printer before, or if you are working on a new type of document, it is important for you first to discuss the following four points, regarding the document's *production stage*, in person or over the telephone:

- Choice of paper
- Use of color
- Layout of page
- Binding and distribution

Choice of paper

The paper that you choose for your document is determined by that document's communication context and message, the document "look" that you want to achieve, and the project budget. Printers offer a variety of paper choices that range in size, texture, color, and cost. The type of paper you choose, probably one of the last decisions you make about the document, may mean the difference between producing an appealing and usable document and one that is not.

Different types of paper exhibit any number of characteristics. For example, paper may appear glossy, shiny, or matte, with a texture that is smooth, rough, thin, or thick. Remember, an important factor to consider when determining the paper for your document is where the document will be read and used. For example, if the document will be read under adverse environmental conditions, such as in a dirty, watery factory, choose a paper that suits this type of environment.

Use of color

Printers follow the same general rules that technical writers follow when color is concerned (refer to the section in Chapter 9 that discusses color as an element of document design). An important principle that printers consider when using color within any document is that *color* attracts the eye before *non-color;* so it is important to use color purposefully and consistently. Printers also understand that readers do not view color in isolation, and color choice should be made with respect to any surrounding color. Because the look and texture of the paper may affect color, examine your color choices on the paper you plan to use.

Once colors are chosen, printers use two color processes in documents: spot color and process color. Printers use *spot color* throughout a document as a second color to complement black. Spot color is used to help format the page and is usually added as an accent in headings, around or in visual aids, or in shading, boxes, and lines. The spot color can be any color, but popular spot colors include gray and blue. *Process color*, achieved by using a four-color separation process, is used to color visual aids, such as illustrations and photographs.

Allison consults with the new printer...

After a search, Allison decides on a printer based on her original criteria of price, turnaround time, service, and comfort. Allison's print representative, Elaine, works with her on all aspects of the newsletter's production stage. Allison provides Elaine with specifications for the SunSpot newsletter that include paper weight and color, size, ink colors, binding/folding specifications, quantity, and delivery information. In the coming months, after these specifications have been set up, Allison will only need to provide Elaine the newsletter "on disk" with a printed "hard" copy, since the printer will have these production specifications on file.

Layout of page

The arrangement and placement of text and visual aids constitute the layout of the page. As a technical writer (with or without the help of a designer), you provide the printer with the layout of your document. Depending on your preferences and those of the printer, the layout you submit to your printer may be camera-ready copy or a rough layout.

Camera-ready copy, as the name suggests, provides the printer with a nearly exact indication of the layout of text and visual aids of the document that you need printed. As a technical writer, you may achieve camera-ready copy through desktop publishing equipment. *Rough layout* provides the printer with a less exact indication of layout. Typically, a rough layout provides the printer with thumbnail sketches and roughly drawn, or pasted-in text and visual aids.

Binding and distribution

After printing is complete, there are two more steps in the production process with which printers are typically concerned: binding and distribution. The type of document you have printed determines whether or not binding is necessary. Typically, the lengthier the document, the more likely binding is necessary.

Saddle-stitching, the most economical type of binding, is used for larger documents, and it is a popular method for binding magazines. With saddle-stitching, a *signature* (a group of pages printed on the same sheet) of folded pages fits onto a V-shaped *saddle*. Staples or thread are then run down the center of the document. This type of binding allows the document, when open, to lie flat and stay open.

Once printing and binding are complete, many printers either arrange for the distribution of the documents or distribute the documents themselves. To make sure that your documents are delivered properly, it is important to consult your printer about distribution information for each document that is printed.

Response Exercise 14.3

Understanding the printer's role in the production of a document means knowing the right questions to ask about making your document successful. To learn more about these design principles, collect four or five magazines of different kinds (home interior, sports, fashion, news, art).

While paging through each magazine, closely examine the features that are important to printers: choice of paper, use of color, layout of page, and binding. Then write down your responses to the following questions about each of these features:

1. What is the look, feel, texture, and thickness of the magazine's *paper?* How does this paper quality differ from the paper used in other types of documents, such as a business letter, textbook, or instruction manual? Why does this difference exist?
2. Find one article where *spot color* is used, and describe how it is used in the article.
3. Find one article where *process color* is used. How would the "look and feel" of the illustration(s) or photograph(s) differ if process color were not used?
4. Open up the magazine to a two-page spread. Create a thumbnail sketch of the *page layout*. What design features make up the layout? Is the layout bysymmetric or asymmetric?
5. Open up the magazine and examine its *binding*. Is the binding saddle-stitch?

Collaborating with either a designer or printer in order to produce a quality document will probably be necessary in your job as a technical writer. Remember, lengthy documents or documents with complicated designs benefit most from the services of a designer. These types of documents, as well as documents that will be distributed widely, will also require the services of a printer.

Developing a good rapport with your designer and printer may mean the difference between a successful and unsuccessful document. Therefore, to create high-quality documents, you need to develop a good working relationship with your designer and printer and know how to communicate effectively with both of these professionals.

Allison is pleased with the printer...

As the print representative, Elaine shows Allison and SunSpot several service courtesies that SunSpot's last print vendor did not. Elaine's company picks up most jobs, which SunSpot considers excellent service. Elaine provides Allison and several other writers at SunSpot with paper sample books to keep. Allison continues to shift the focus of the SunSpot newsletter to include the corporation's "new millennium" theme. Overall, both newsletter readers and SunSpot's management seem pleased with the way Allison is handling this newsletter project.

Chapter 14 Task Exercise

Knowing the printer's role in document production will make your collaboration with this important colleague more efficient and productive. To better identify the printer's tasks and responsibilities, visit a local printing firm in your area. Schedule your visit so that you can ask the printer questions about his or her craft. To understand the types of tasks performed by this printer, write down the printer's responses to the following questions:

1. What are the ways in which your firm sells its services to prospective customers? Advertising in newspapers or magazines? Word-of-mouth?
2. When a prospective customer—perhaps a technical writer who works for a company or organization in the area—contacts you as a possible printer for their company or organization, what types of information do you discuss with him or her? Do you discuss basic printing services, cost, or incentives offered to preferred customers?
3. As a new customer, if I needed 150 copies printed of a brochure, what would we need to discuss in order to print that document?

Next, request that the printer give you a tour of the printing facilities. Then you may ask the printer specific questions about the equipment used to perform certain tasks, the number of personnel required to run the equipment, and the tasks those personnel must perform. Be sure to schedule this tour when it is convenient for both you and the printing firm employees.

Part IV

Creating Readable Products

Chapter 15

Understanding Expectation and Interpretation in Reading

To produce quality documents, you should understand not only how to improve your writing process, but also what strategies your readers use to understand those documents.

To begin the earliest planning stage of writing any document, a technical writer realizes that identifying and understanding that document's communication context—its situation, purpose, and audience—is important. Successful technical writers also recognize that understanding *how readers read* is just as important for creating effective documents. Specifically, writers must know the similar ways in which their readers read and the comparable strategies that readers use to understand a document or text. While a given document's audience may be diverse in terms of background, education, and expertise, technical writers can count on the fact that important similarities in reading process and strategies exist among readers.

Understanding the features of the reading process is important, yet to be a successful technical writer, you must know how to use this knowledge to create more readable documents. This chapter covers not only how readers read, but also how writers can learn from this and include particular elements that make documents more usable and reader friendly.

While we discuss how technical writers use information about the reading process to create better documents, we also introduce reading strategies from which everyone can benefit. Sandwiched between discussions about the reading process and creating better documents is a discussion about reading strategies. That is, we identify the ways that all readers—including you—can make reading easier and more effective.

While our considerations about reading are relatively broad, this chapter gives you basic, useful information. After having read this chapter, you should be able to respond to the following questions about reading:

- What do you need to know about *how readers read?*
- What *strategies* can help you read more effectively?
- What *elements* can make the documents you write easier to read?

Since both writers and readers benefit from learning more about the reading process, this chapter presents a variety of information. The points we make and the suggestions we give are useful for many different types of readers and writers.

What do you need to know about how readers read?

Understanding the differences among a document's readers—their education, expertise, prior knowledge, and expectations—is important for effectively communicating to a range of audience types. However, recognizing the similarities among readers in terms of how they may approach a document, construct meanings from it, and understand a document's message, is important for communicating more efficiently and effectively.

While reading is a highly dynamic activity, readers still tend to follow certain methods during this process. First, readers rely heavily on *prior knowledge*, or what they already know about a subject, to guide and inform their understanding of the document. If readers know very little about a subject, reading a highly specialized document, such as an article from a professional journal, will make little sense to them. As a writer, if you know your readers may have little prior knowledge about a subject, begin your discussion of that subject at a basic level, so that readers can begin to build their knowledge.

Second, readers demand texts that *make sense* to them; that is, texts that are purposeful, well organized, and present content in a logical manner. If readers are confronted with a text that is unorganized or presents its message illogically, readers tend to tune out or dismiss the document altogether. As a writer, work with what you know about your audience's needs to provide them with a suitable document in terms of purpose, organization, and content.

Third, readers read for certain *purposes,* whether they consciously realize it or not. In the following discussion, you will learn about three different purposes that readers commonly share for reading texts. While each of these purposes is described separately, a reader usually has multiple purposes for reading a single document. Therefore, when reading a document, a reader may first *skim* the text for its main points, closely read the text to *learn* about something, and also read the text to *do* something—to perform some process or task (Redish, 1988).

Skim

While readers skim different types of documents in different ways, their general reason for doing so is to discover whether or not it is necessary to read the text thoroughly. Depending upon the type of document, readers may skim by quickly reading headings and subheadings, glancing through an executive summary or abstract, or quickly reading the first several paragraphs of each section of the document.

By skimming a text, readers decide whether they want or need to read it more thoroughly. Also, readers determine on which part of the text to concentrate, skip, or read first. Depending on your familiarity with the type of text that you are skimming—for example, a memo, report, or set of equipment specs—you probably skim in different ways.

Learn

By reading to learn, readers are searching to understand meaningful information, concepts, and ideas. More experienced readers tend to read to learn more successfully than those who are less experienced. As a result of more experience, readers pick up reading strategies that enable them to learn more quickly and completely. In the next section, you will learn several strategies that may help you to increase your ability to read to learn.

As a result of reading to learn, readers may make decisions, solve problems, or take action based on their new knowledge. Readers may also determine that they need to read more about a subject before moving forward in these directions. Nearly everyone—from students to business professionals—spends a great deal of time reading to learn.

Do

By reading to do, readers use the information they have read or are reading to perform a task or process. At some point, all readers read to do, whether they read to assemble a child's bicycle, perform an experiment in a lab, or operate a new software application.

As with other purposes for reading, certain texts allow audiences to read to do more easily than others. By reading to do, readers may perform a range of tasks. Readers may also determine that they need to read more about a subject before moving forward. In the next section, you will learn several strategies that may increase your ability to read to do.

Response Exercise 15.1

In the preceding section, we discussed what you need to know about how readers read—the processes they move through, and the purposes behind the reading that they perform. To better understand these important concepts regarding reading, write down your responses to the following questions:

1. What three elements do you need to understand, as a technical writer, about the ways in which your audience reads?
2. What are the three purposes that readers typically have for reading?
3. After identifying each purpose, briefly describe why readers perform each, and what is important about purpose. For example, why do readers skim a document? What is important about that process? What does it help readers remember/perform/change?

What strategies help readers read more effectively?

Technical writers read their own documents and documents written by colleagues. Several reading strategies can help you to read more effectively, and to become a better reader, editor, and colleague. Understanding the strategies described below benefits technical writers by leading to an understanding of what their own readers can do to read more effectively.

The following strategies—manipulating your environment, recognizing genres, pausing to question, and remembering to reflect—are practical for anyone to use. Reading with these strategies in mind helps you to write more effectively—whether you are reading to revise your own document or to edit a colleague's document.

Manipulating your environment

An important but often overlooked strategy for aiding reading comprehension is manipulating your reading environment. As a reader in the workplace, you may read in a variety of different environments—noisy, distracting, or quiet. Obviously, if your reading environment is noisy or overly distracting, you may not be paying close attention to the details and ideas presented in the document.

One way to overcome these reading problems with your environment is to (a) identify the distractions, (b) eliminate the distractions or move to another area, and (c) read exclusively in an environment that is suitable for reading.

Recognizing genres

One strategy that many readers tend to implement without even realizing it is the ability to use what they know about the genre—the type of document—to help shape the reading and understanding of that document. Knowing the genre allows you to move to the areas of the text that are of most interest and importance to you, and to skim the text more quickly and effectively.

One important method you can use to improve your ability to recognize different genres is to read more often. Read as many different types of documents as possible, and always consciously attempt to identify the qualities that characterize each type of document. (Note: See the section in Chapter 3 that identifies and describes a number of different workplace documents.)

Pausing to question

One of the best methods that you can use to strengthen your reading comprehension is to stop periodically during the reading process to question yourself about the main and supporting points that the document presents. Typically, readers question themselves about the material that they have just read in order to ensure that they fully understand the concepts and ideas being introduced.

The questions that you ask yourself about a document depend, of course, on the document itself. However, the questions listed below are basic enough to be applied to nearly any type of document:

Questions readers use to *question* the text

- What is the topic of this document/section/paragraph? What is the claim being made?
- What facts, anecdotes, or other details help to illustrate the topic? What evidence helps to support the claim? How persuasive is this evidence?

- Besides the topic, or main idea, what other supporting topics are discussed?
- How are these topics organized and presented? How does this organization affect my acceptance or understanding of them?

Remembering to reflect

Another useful method for improving your reading, and one that is closely related to questioning the text, is the ability to reflect on what is being presented. Readers reflect on the material that they have read not simply to understand it, but also to interpret, analyze, and critique this information. Readers use several different methods to help them reflect—such as careful note-taking, synthesis, and analysis.

Careful *note-taking* on your reading material should occur while you read. Pausing periodically to take notes about important claims or ideas, relevant details, or questions about unclear concepts is a valuable practice. The act of note-taking will help you to reflect about the content of the document, and the notes you keep will serve as an archive that you can refer to in the future.

Note-taking, in many ways, helps to yield *synthesis.* Synthesis is the ability to take what are at first seemingly disparate points and assemble them into a meaningful, new whole. Synthesis may occur during your reading, or it may take place after you have read a document in its entirety. Synthesis is an important part of reading comprehension, and if you cannot synthesize, you may need to re-read.

Analysis moves synthesis one step further, prompting a reader to carefully scrutinize the points being made, and how they are synthesized. Analysis allows for a more in-depth examination of why a certain concept is presented in a particular context, what characterizes the concept, and what about the concept is being valued. After readers analyze a passage or a whole text, they take a position regarding the document, either generally agreeing or disagreeing with its message.

The most useful and constructive analysis of a document is prepared only after a thorough analysis of the document's points. The questions you ask yourself to perform this type of analysis depend, of course, on the document itself. However, the questions below are basic enough to be applied to nearly any text.

Questions readers use to *analyze* a text

- What points are being presented? What claims are being made?
- In each case, what is being valued?
- What points/evidence/facts support the claim? Do I find this support adequate? Convincing? Off-putting?
- Do I generally agree or disagree with the points raised in the text? Why?

Learning useful reading strategies enables you to become more aware of your own reading practices. More importantly, this awareness allows you to begin to revise your own reading practices to make them more effective and efficient. In the following section, we build on this discussion of reading strategies to identify and describe several ways that writers—through the ways in which they create texts—can enable readers to read more efficiently and effectively.

Response Exercise 15.2

The strategies discussed in the preceding section—manipulating your environment, recognizing genres, pausing to question, and remembering to reflect—are practical for any reader to use. Reading with these strategies in mind helps you to write more effectively. To better understand how to use these strategies as a writer, write down your responses to the following questions:

1. What are the four strategies that you can use to read more effectively? In four or five sentences, define each strategy.
2. Beyond the general benefit of improving reading comprehension, each strategy has a *specific* benefit. What is the benefit of each strategy?
3. The last strategy, remembering to reflect, includes three processes that you can follow to help you better reflect. What are these three processes?

Choose a document that you have never read before: a letter, brochure, manual, or other workplace document. Using the reading strategies described in this section, read the document, and write down your responses to the following questions:

4. Identify your purpose(s) for reading the document. Did you skim, read to learn, or read to do? Did the type of document you chose affect your reading purpose(s)?
5. Describe the benefits of using the four reading strategies. Did your understanding of the material come quickly? If anything, what hampered your understanding of the material?
6. Will you use any or all of these reading strategies again? Why or why not?

What elements make your documents easier to read?

Several strategies can help you, as a technical writer, create documents that readers can read more effectively. Many of the strategies used to produce reader-friendly documents result from what we know about how readers read, and the strategies that effective readers use to read successfully. Document strategies like these can be implemented at any point during the writing process, from the planning through editing stages; however, including these elements relatively early on in the writing process makes them easier to implement.

The following elements are categorized into three areas: (a) using genre expectations, (b) creating logical organization, and (c) working with readers' prior knowledge. By thinking about and using these elements within your documents, you enable your readers to better concentrate on the important message that you are communicating.

Using genre expectations

A specific genre—or type of document—contains certain elements that help readers to characterize and recognize it, and readers look for these elements in particular places in the document. For example, readers who use an instruction manual look for certain elements in that manual: text written in complete, concise sentences; illustrations that are clearly labeled; and text organized in a step-by-by manner. Readers of instruction manuals *expect* elements such as these to appear, and when they do not appear in the document, readers experience more difficulty reading and understanding.

The questions that you ask yourself about a document's genre expectations depend upon the type of document you are writing. However, the basic questions listed below are applicable to nearly any text:

Questions writers ask themselves about using *genre expectations*

1. What genre—or type of document—am I creating? What is this document's communication context?
2. What elements typically characterize this type of document?
3. What elements will my readers look for in the document? Where are these elements usually placed?
4. What other methods, related to genre expectations, can I use to help guide my reader through the document?

Remember to help your readers read more effectively by using their expectations about genres. If possible, ask a colleague to read your document for its effective use of genre expectation. To create the best possible document, ask that colleague to suggest ways to use readers' genre expectations more fruitfully.

Creating logical organization

Readers react favorably to documents that are well organized and exhibit a logical structure. Once again, as a writer, you must understand the document's communication context—including your readers' expectations—in order to best organize that document to suit the needs of your audience.

Once you have determined how to successfully organize the main and supporting points of your document, you must employ a variety of well-placed cues to guide your readers through the organization of that document. Depending on the type of document, you ask yourself different questions about the ways to best organize that document. However, the questions below are useful, and are applicable to nearly any text:

Questions writers ask themselves about creating *logical organization*

- What organization strategy is best for these readers? What strategy conveys my message most clearly and persuasively?
- What cues can I use to guide my reader through this organization? Should I use a table of contents, headings, subheadings, forecasting statements, clearly written topic sentences, visual aids, color, lines, or other formatting devices?

Working with prior knowledge

Readers understand new information more readily and completely when you introduce it by first providing them with more familiar information. This *given-new* strategy allows readers to base their understanding of the new material on information they already understand. This strategy is useful for organizing an entire document, a section of the document, a paragraph, or even a sentence.

The following paragraph is excerpted from a textbook written for high school science students. Prior to this section, the text has already introduced and defined *hormones*, substances humans produce that affect growth, metabolism, and digestive functions. The new subject being introduced in this section is endocrine glands.

Figure 15.1 Textbook excerpt

> Humans produce hormones and distribute them through the body in a number of ways. One way in which hormones are distributed—or secreted—is directly into the blood stream, through the endocrine glands. Endocrine glands secrete a number of different hormones into the body which affect a number of different functions, from your physical development to the balance of minerals in your body.

Notice how the excerpt begins with one sentence reminding readers about hormones: "Humans produce hormones and…." This sentence triggers readers' prior knowledge about hormones—their function in the body. The second sentence ("One way in which hormones are distributed…") builds on this prior knowledge by linking the idea of hormone distribution to the new topic being introduced—endocrine glands. This excerpt concludes with a one-sentence definition of this new topic. This given-new strategy successfully builds on students' prior knowledge about hormones to introduce them to the function of endocrine glands.

While this excerpt is taken from a textbook, the strategies used in it are applicable to nearly any type of document. Using what you know about your audience and the type of document that you are creating, you should work with your readers' prior knowledge. To accomplish this, begin by asking yourself the following questions:

Questions writers ask themselves about *prior knowledge*

- What do my readers already know about this subject? What information are they lacking about this subject?
- Specifically, what new information should be presented first, in order to provide them with a context with which to read the new information?
- Should this given-new strategy be used document-wide, or just within a section, paragraph, or sentence?

Chapter 15 Task Exercise

By implementing three factors related to how readers read—using genre expectations, creating logical organization, and working with readers' prior knowledge—you can create more user-friendly documents. In this section, we suggest specific ways to include these factors in the documents that you create.

Access and read through a typical workplace document; you may choose correspondence, a manual, report, or brochure. Using what you know about how readers read, write down your responses to the following questions:

1. What type of document have you chosen? Describe, as much as possible, the communication context for this document. That is, what is the document's situation, purpose, and audience?
2. What do you know about the intended audience of this document? For example, what type of education, expertise, or prior knowledge do they have? What are their values, interests, and investments in the topic of this document?

3. What *genre expectations* might this audience have regarding this document—that is, what elements do readers expect to find in the document? Are these elements included? What other elements would you include to work with this audience's genre expectations?
4. What type of *organization* this document exhibit? Is this organization suitable? What elements guide the readers through this organization? What other elements would you include to guide the audience?
5. How does the document work with the readers' *prior knowledge* about this document's subject? Identify at least one document-wide given-new strategy and one given-new strategy used in a paragraph.

As a technical writer, you must know the ways in which your audience reads, as well as the similar strategies that readers use to understand a document or a text. While you may be writing for an audience diverse in background, education, and expertise, as a technical writer, you can include certain elements in the document to help your audience read and understand its message. Therefore, understanding the features of the reading process enables you to create more usable and reader-friendly documents.

Chapter 16

Creating Logical and Coherent Documents

To become a good technical writer, you should learn to create documents that readers find useful and interesting to read. Typically, documents that feature these qualities are also logical and coherent.

In these final two chapters, we focus on ways that you can further sharpen your writing skills. To do this, we suggest specific strategies that are applicable for use in a wide range of technical communication—from correspondence and reports to proposals and Web sites. In this chapter, we identify and describe several techniques that you can use to produce documents that are *logical* and *coherent*. The techniques you learn about here can be executed during any stage of the writing process, from planning through editing.

Logic and coherence are two important qualities of successful technical communication. Most readers and writers quickly come to realize why logic and coherence are important. However, writers—particularly those who are less experienced—find that creating documents that exhibit these qualities is a difficult undertaking. For example, many writers can only begin to revise for logic and coherence after they have created a nearly complete draft of the document, while other writers can only revise for these qualities after being given specific suggestions for doing so by an editor or colleague.

In other words, be ready to work hard on making these strategies work in your documents, and do not expect to be able to execute them flawlessly the first time you use them. To begin understanding how to create logical and coherent documents, you should know the answers to the following questions:

- What document-wide *organization strategy* should you choose?
- What *visual cues* can you include to guide your reader through the document?
- What *textual cues* can you include to guide your reader through the document?

By consciously implementing the strategies we discuss in this chapter, you should begin to create documents that are both logical and coherent, while improving both your written products and your communication skills.

What document-wide organization strategy should you use?

An organization strategy is the way you choose to arrange the information in your document. An organizing strategy must be chosen purposefully for each document that you write. While a particular strategy works well in one type of document, that same organizing strategy may fail in another type of document. There are several different types of organizing strategies to choose from.

To choose a strategy that is right for your document, you must ask yourself these questions about the document itself and about organization. (Note: For more information about organization strategies, including types of strategies to choose, see Chapter 6.)

Questions writers ask themselves about choosing *organization strategies*

1. What is my document's communication context—that is, what is its situation, purpose, and audience?
2. Given this context, what type of organization strategy seems most appropriate?
3. In what ways is the strategy suitable/not suitable for my situation? My purpose? My audience?
4. What strategies have succeeded in past documents? Would those strategies be useful and appropriate here?

You may decide on an organization strategy early in the writing process of a document, say, in its planning stage. However, as you write the document you may find that another strategy is more useful, or you may discover that using a different strategy in another section of the document is suitable. As with most document decisions, choosing an organization strategy should be a thoughtful, purposeful one.

Choosing an organization strategy and using it in your document is a necessary first step toward creating a logical and coherent document. In the following two sections, we identify and describe specific ways that you can improve the logic and coherence of your document once the organization strategy is in place. In the next section, we discuss what visual cues you can use in your document to help guide your reader.

Response Exercise 16.1

In the preceding section, we discussed the importance of devising an organization strategy for your document, and how this choice affects your document's logic and coherence. To better understand these important concepts, write down your responses to the following questions:

1. What do you believe is the relationship between choosing an organization strategy and creating a document that is logical and coherent? In what ways does one inform or impact the other, and vice versa?
2. A document's intended readers influence the organization strategy that you choose. As a technical writer, what methods can you use to find out information about this audience? (Note: You may want to review the discussion in Chapter 5 about understanding audience as part of the communication context.)
3. In Chapter 6, you learned about several different organizational strategies. Describe the differences of each of the three strategies listed below, and indicate the purposes for using each in a document. For example, what elements does a cause and effect strategy contain that make it suitable for a particular type of document?

    - Cause and effect
    - Chronological
    - Compare/contrast

What visual cues help to guide your reader?

After you have chosen an organization strategy for your document, use visual cues—non-textual elements that purposefully and consistently catch a reader's eye—to guide your reader through that document. Visual cues help to reveal the logic and coherence of a document's organization. Readers who encounter suitably chosen and well-placed visual cues can more easily find their way through a document, and are thus more likely to grasp and understand the message the document communicates. The two types of visual cues we discuss are a collection of formatting devices (lines, boxes, and color) and clip art.

Lines, boxes, and color

As visual cues, lines, boxes, and color may be used either as separate elements or in combination with one another in a document. The uses of these particular visual cues are many, and listed below in Figure 16.1 are several different ways to incorporate them effectively into your document.

Figure 16.1 Uses of lines, boxes, and color

Lines—bold, wide, thin, dotted, dashed
• To divide chapters, sections, subsections of a document • To separate one type of information—a definition, exercise, or summary—from the main body of the document • To emphasize a phrase or a passage (underlining) • To separate information within a table, or other visual aid
Boxes—three-dimensional, shadowed, plain
• To separate one type of information—a definition, exercise, or summary—from the main body of the document • To call out or emphasize one paragraph or passage
Color—accompanying text, visual aids, alone
• To divide chapters, sections, subsections of a document (used with headings or alone) • To emphasize certain sections or passages • To organize information in a visual aid, such as a pie chart, bar graph, or line graph

While lines, boxes, and color can be used within documents in a multitude of ways, these visual cues only have two basic purposes—to separate information and to emphasize information. Visual cues separate information that is topically different from the main body of information, or information that must be used in a different way.

Visual cues can emphasize certain information, such as a section, paragraph, or brief passage. Color can be used to draw attention to an important safety issue, while other visual cues, like boxes, can be used to emphasize an important one-sentence definition. Remember, visual cues, such as lines, boxes, and color, are more effective when they are used consistently throughout the document.

Clip art

Clip art is a picture, icon, or other graphic that can be inserted into a document. Clip art can be a picture that depicts a recognizable scene, object, or image, such as the one in Figure 16.2.

Figure 16.2 A picture

Clip art can be an icon, which is a recognizable symbol that can be used to give readers reference points within a text For example, the stop sign icon in Figure 16.3 can be used in an instruction manual to grab readers' attention, and to emphasize the important safety information to which it is attached.

Figure 16.3 A stop sign icon

Clip art can be visuals whose main purpose is to be decorative or aesthetically pleasing. Visuals such as the one in Figure 16.4, while decorative, are still useful in that they may serve to separate chapters of a text, or sections of report in a pleasing manner.

Figure 16.4 A decorative, aesthetically pleasing clip art item

As a writer, you can use individual clip art items alone, or in combination with other visual cues to guide your reader through the document. Visual cues, whether they are lines, boxes, color, or clip art, are valuable ways to reveal your document's organization strategy to the audience, and enable them to better understand the message your document communicates.

Response Exercise 16.2

Visual cues are non-textual elements that purposefully and consistently catch a reader's eye, and guide that reader through the document. Choose a document that you have never read before. You may choose a letter, brochure, manual, or other workplace document. To better understand how visual cues work in documents, read the document that you have chosen, and write down your responses to the following questions:

1. Identify and describe the visual cues writers may use to help guide their readers through a document.
2. What visual cues are included in your document?
3. Is one type of visual cue used more often than others? If so, in what specific ways does it guide the reader? If not, what combination of cues are used? In what specific ways does this combination guide the reader?
4. As a reader, what visual cue(s) did you find most effective? What about the placement/format/purpose of the cues made them particularly useful?

5. What visual cue(s) did you find least effective? What about the placement/format/purpose of the cues made them a hindrance?
6. Did another type of visual cue need to be included? If so, where?

What textual cues help to guide your reader?

To guide readers through a document, good technical writers use a combination of both visual and textual cues. While visual cues are non-textual elements, such as clip art, lines, boxes, or color, textual cues use the written word as a device to guide readers.

Like visual cues, textual cues help to reveal the logic and coherence of a document's organization. Readers who encounter suitably chosen and well-placed textual cues can more easily find their way through a document, and are thus more likely to understand the message the document communicates. The types of textual cues we discuss are even more effective when used in combination with visual cues.

Headings and subheadings

Headings, which are titles of the main sections of a document, and subheadings, which are titles of subsections of a document, are common and useful types of textual cues. Headings and subheadings allow readers to distinguish one section of the document from the next, and to show the relationship between, or hierarchy of, those document sections.

While considered textual cues, headings and subheadings are useful only when well formatted and carefully designed. Headings must exhibit an appearance that easily differentiates them from the surrounding text while remaining consistent with each other in font size and style. Likewise, subheadings must be formatted consistently to distinguish them from both the surrounding text and the headings. (Note: It is best to assign to headings and subheadings styles that complement each other but cannot be confused with each other.)

Purpose statement

A purpose statement is actually a collection of sentences that identifies the *project problem* that the document attempts to solve, the *tasks* or work completed to solve that problem, and the *rhetorical purpose* or purposes for writing the document. A purpose statement containing these three elements can be included in correspondence, informational material, or virtually any type of report. Figure 16.5 contains a purpose statement excerpted from a recommendation report.

Figure 16.5 Recommendation report purpose statement

> To complete this project, we were asked to test the efficiency and cost-effectiveness of two similar equipment models: Model SRT and Model 198-680. We tested each model according to the parameters our team and members of your organization devised. This report contains not only a description of that work, but also the test results and our recommendations.

Experienced readers expect purpose statements such as this, particularly in lengthier documents such as final reports. A purpose statement reminds readers about the document's context and background. Readers become focused by examining the purpose statement, and it helps to add coherence to your document.

Forecasting statements

Forecasting statements should be concise and complete sentences that are placed consistently throughout the document, from beginning to end. Forecasting statements can forecast the main points of an entire document, one section, or one paragraph. Forecasting, whether it is included in a sentence, a list, or a series of sentences, projects the order of important information in the document. Forecasting guides the reader from point to point, and helps the reader to anticipate what information comes next.

Topic sentences

Two textual devices that work well to help readers on the sentence- and paragraph-levels are topic sentences and transitions. Topic sentences present the main idea of the paragraph and are usually the first sentence of that paragraph. Complete and concise topic sentences allow readers to understand the scope or focus of that particular paragraph. Topic sentences do not forecast points or claims the way forecast statements do. Rather, topic sentences *introduce* the topic of a paragraph.

Transitions

Transitions are words or phrases that help to connect concepts or ideas. Transitions that are appropriately chosen and well placed help to reveal to the reader the logical connections among ideas in a document. Transitions, such as those listed in Figure 16.6, can signal a wide range of relationships among ideas.

Figure 16.6 A list of typical transition words and phrases

Purpose	*Transitions*
Compare	likewise, similarly
Contrast	however, yet, but, on the other hand, nonetheless
Outcome	therefore, so, accordingly
In brief	finally, in summary, in conclusion

Both transitions and topic sentences are textual devices that shuttle the reader from idea to idea, in a logical and coherent fashion. When textual devices like these are missing, or when they are inappropriately placed, readers become confused and may choose to ignore that section of your document, or the document altogether.

Chapter 16 Task Exercise

Textual cues use the written word as a device to guide readers through the document. Like visual cues, textual cues help to reveal the coherence of a document's organization.

Re-read the document you used in Response Exercise 16.3. To better understand how visual cues *and* textual cues work in documents, write down your responses to the following questions about that document:

1. Identify and describe the textual cues writers may choose from to help guide their readers through a document.
2. What textual cues are included in your document?
3. Is one type of textual cue used more often than others? If so, in what specific ways does it guide the reader? If not, what combination of cues are used? In what specific ways does this combination guide the reader?
4. As a reader, what textual cue(s) did you find most effective? What about the placement/format/purpose of the cue(s) made it particularly useful?
5. What textual cue(s) did you find least effective? What about the placement/format/purpose of the cue(s) made it a hindrance?
6. Did another type of textual cue need to be included? If so, where?

Next, review your responses to the questions in Response Exercise 16.3, and respond to the following questions:

7. Identify and describe one instance in which both visual and textual cues worked together, and complemented one another, in the document.

8. Why is this combination more/less effective than if the cues were used individually? What changes, if any, would you make to this particular combination?
9. What visual cue would you include to complement a textual cue? What textual cue would you use to complement a visual cue?

Specific strategies—applicable for use in a wide range of documents—can help you to make the products you create more logical and coherent. While many writers find that producing logical and coherent documents is a difficult undertaking, writers know that these characteristics help to define successful technical communication. With practice, your documents can be both logical and coherent.

Chapter 17

Writing Readable Sentences

To become a good technical writer who produces appealing and usable documents, you should develop a prose style that is clear, concise, and courteous.

In this chapter, we identify and describe several techniques that you can use to improve the readability of your writing or *prose.* Readability refers to the way in which your sentences read. For instance, your prose may read easily, poorly, or be nearly incomprehensible. You can improve the readability of your sentences largely by revising the way they are put together.

With all types of documents, your prose should exhibit three qualities: clarity, conciseness, and courtesy. When your prose is *clear,* its sentence structure assists in conveying meaning, rather than hindering or detracting from it. When your prose is *concise,* its sentences do not contain an overabundance of words, yet the intended meaning is readily communicated. When your prose is *courteous,* its sentences exude a tone that is professional and respectful of your readers.

When your prose exhibits clarity, conciseness, and courtesy, the message of your document is more easily and effectively communicated to your audience. While the strategies for improving your prose style are practical and relatively straightforward, putting them into practice is challenging. Many times, writers have difficulties with one particular aspect of their prose (such as clarity) and are hard-pressed to completely remedy it. With practice, however, you should be able to improve your prose style and improve the effectiveness of all of your documents.

To begin discovering ways to improve the readability of your prose, you should pay attention to issues of clarity, conciseness, and courtesy. To accomplish this, you should know the answers to the following questions:

- How can you develop *clear* prose?
- How can you develop *concise* prose?
- How can you develop *courteous* prose?

As a writer, knowing basic information about improving aspects of your prose style enables you to produce better-quality documents. By consciously implementing these strategies and always striving to make your style as readable as possible, you can improve your writing and communication skills.

How can you develop clear prose?

In Chapter 3, we discuss those qualities that technical writers use as criteria to measure effective technical communication. When wielded correctly, these elements—an attention to communication context, completeness in terms of content, purposeful organization, effective prose style, and attention to design—characterize successful communication. Notice that effective, readable prose is only *one* of the five qualities that characterize effective communication. Therefore, as a technical writer, you should work to develop all of these qualities in the materials that you create.

Developing clear prose that assists rather than detracts from the meanings you intend to communicate is difficult. However, by paying attention to your documents at the sentence level and following the strategies suggested in this chapter, you can begin to improve the readability of your prose. Specifically, to improve your prose's *clarity*, attend to three elements: voice, expletives, and word choice.

Choosing between active and passive voice

Writers have to choose between active voice and passive voice all of the time. However, few beginning writers can articulate the different purposes of each voice. Deciding between active and passive voice should always be a purposeful choice for technical writers. First, you should understand how each voice is constructed in a sentence.

In the *active voice*, the subject performs the action named by the verb:

Example: *Claire's team* cut the budget twenty-five percent.

In the *passive voice*, the subject of the verb receives the action:

Example: The budget was cut twenty-five percent (by *Claire's team*).

Notice that in the passive voice example, the subject of the sentence, "Claire's team," is in parentheses. The parentheses illustrate that in passive voice, the subject is not explicitly stated. Therefore, you can read the passive voice sentence without reading its subject, "Claire's team."

To further illustrate the purpose of each voice, read the two passages in Figure 17.1.

Figure 17.1 Passages illustrating active voice and passive voice.

Active voice: After meeting for nearly sixteen hours total, the team cuts Fiscal Company's budget twenty-five percent. Claire's team cuts spending in three areas—research and development, personnel, and technology—in an attempt to propel Fiscal out of the red and into the black.

Passive voice: Fiscal's budget is cut twenty-five percent, after meeting for nearly sixteen hours total. Spending is cut in three areas—research and development, personnel, and technology—in an attempt to propel Fiscal out of the red and into the black.

Figure 17.1 shows that active voice implicates the subject (Claire's team) in its activity—cutting Fiscal's budget. The passive voice, however, removes this subject from the budget-cutting action.

Avoiding unnecessary words

To develop clear prose, you want to avoid cluttering sentences with unnecessary words, or *expletives*. Common offenders are the constructions *there is/are/was/were* and *it is/was*. By eliminating these constructions, particularly at the beginnings of sentences, you are forced to revise and tighten the construction of the entire sentence. To illustrate this, note the differences in clarity produced by first including and then eliminating the unnecessary *there* and *it* constructions in Figure 17.2.

Figure 17.2 Passages illustrating expletive use

With unnecessary constructions: There is a real hesitancy, on the part of nearly everyone attending the meeting, to cut the budget. It is necessary, though, that the team alter the budget. During last night's meeting, there was a lot of discussion among the team members about which cuts to make.

Without unnecessary constructions: Nearly everyone attending the meeting hesitates to cut the budget. The team must necessarily alter the budget. During last night's meeting, team members discussed which cuts to make.

Polishing word choice

As a technical writer, you choose your words by using precise word meanings and concrete language. Using *precise word meanings* involves knowing the difference between a word's denotative and connotative meaning. Denotative meaning is the more literal, dictionary definition, while the connotative meaning is the more implicit definition (Figure 17.3).

Figure 17.3 Comparing denotative and connotative meanings

The team decides which *budget cuts* to make.	*Denotative* • Budget reduction	*Connotative* • Budget reduction, a necessary evil, may result in lower quality of life for workers.

Polishing word choice also involves using *concrete language:* specific and precise words and phrases meant to define, explain, or describe a particular topic. Figure 17.4 compares concrete language to vague, less direct language.

Figure 17.4 Comparing concrete and less direct language

Concrete language	*Less direct language*
The team's meeting lasted for three hours—from 3 p.m. until 6 p.m. Five of the seven team members attended the meeting. The team cut the budget by 5.60%.	The team's meeting lasted a really long time. Most of the team members attended the meeting. Finally, the team cut the budget.

Response Exercise 17.1

To improve your prose's *clarity,* you should attend to three elements: choosing between active and passive voice, avoiding unnecessary words, and polishing word choice. Using what you have learned about these elements, revise the passage below for clarity.

The following passage is an excerpt from the *Fiscal Workers Newsletter,* a bi-weekly publication written, printed, and distributed by (and to) Fiscal workers and their families. This excerpt is a rough draft and needs to be revised for clarity.

> ***Excerpt:*** There has been a lot of talk lately about the recent budget cuts at Fiscal. It is pretty clear that a lot of people in our town—many of whom work for Fiscal—feel these changes. After talking with lots of these people in town, there are three main ways that the budget cuts have made a difference in the lives of townspeople: sense of hopelessness about the future, feeling of betrayal, and anger.

How can you develop concise prose?

Many busy professionals state that one of the most desirable qualities for prose to possess is conciseness. A document that exhibits concise prose enables its audience to understand its message accurately and efficiently. That is, the meaning of the document is conveyed without the use of extraneous words and phrases, while the message is communicated completely.

Both experienced and novice writers often have difficulty revising their prose for conciseness. In the following discussion, we identify two related processes—combining sentences and eliminating empty phrases—that may help you to develop more concise prose.

Combining sentences

To create accurate, complete, *and* concise sentences, you should avoid using unnecessary constructions. These include clauses or sentences that can be significantly revised or eliminated without altering the intended meaning of the sentence. To develop concise prose, remember to search for these unnecessary constructions, and omit them by combining sentences.

When two related sentences contain extraneous words, you can make them more concise by identifying the most significant words or clauses in each of the sentences, combining them, and creating one complete, accurate, and concise sentence. *Sentence combining*, when performed correctly, is one of the most useful ways to create a more concise prose style. By comparing the passages in Figure 17.5, you can see the effectiveness of sentence combining.

Figure 17.5 Illustrating the effects of combining sentences

Before: Fiscal Corporation has managed real estate ventures all over the United States. During the last two years, Fiscal Corporation has reported net losses of nearly 2.5 million dollars.

After: During the last two years, Fiscal Corporation—an organization that has managed real estate ventures all over the United States—has reported net losses of nearly 2.5 million dollars.

In Figure 17.5, repeating "Fiscal Corporation" twice is an unnecessary construction. Therefore, sentence one is turned into a definition, and is set off by dashes. In this case, sentence combining took two sentences and revised them into one, more concise sentence, while preserving the meaning.

Eliminating empty phrases

Inexperienced writers often have the most difficulty avoiding the use of empty words or phrases in their writing. Empty phrases, such as "What I am trying to say is..." or "The purpose for writing this report is that..." generally do not add any new meaning to the passage or document. Readers may find that these empty phrases hinder clarity, conciseness, and comprehension. Therefore, try to avoid empty phrases, such as those listed in Figure 17.6.

Figure 17.6 Empty phrases (Houp, et al., 100)

To the extent that
With reference to
In connection with
Relative to
With regard to
Is already stated
In view of
Inasmuch as
With your permission
Hence
As a matter of fact

Response Exercise 17.2

In the preceding section, we identified two related processes—combining sentences and eliminating empty phrases—that should help you to create more *concise* prose. Using what you have learned about these two processes, revise the following passage for conciseness *and* clarity. (Note: While this section focuses on creating concise prose, you should try to revise the passage for both conciseness and clarity.)

Once again, the following passage is an excerpt from the *Fiscal Workers Newsletter*. This excerpt is a rough draft and needs to be revised for clarity and conciseness.

> ***Excerpt:*** It is likely that the recent budget cuts are just one more of Fiscal's economic restructuring processes. As a matter of fact, Fiscal probably has many more ideas about what kinds of restructuring should happen in the future. With regard to the cuts or layoffs that may be in our future, it is difficult—as a Fiscal employee—to make any personal, economic plans for the future. My economic plans for the future have been put on hold.

How can you develop courteous prose?

In many ways, developing concise and clear prose involves revising the structure of your sentences. But producing prose that is courteous—exudes professionalism, and shows respect toward your audience—cannot be developed simply by changing the structure of your sentences. Generally, if you identify the values that your readers share, and show them that you respect those values, your prose is courteous.

Courteous prose is not simply a professional gesture, but also it helps to establish and maintain good working relationships with your readers and to improve the success of your communication. To develop courteous prose, avoid biased or inappropriate language, sexist language, and jargon.

Biased or inappropriate language

Eliminating biased or inappropriate language is akin to maintaining your professionalism. Courteous prose is an important persuasive factor, and if you are discourteous, your document may lose some or all of its persuasive power. To illustrate the importance of courteous prose, examine the two passages in Figure 17.7; one exhibits language that is biased and inappropriate, while the other is courteous and professional.

Figure 17.7 Biased and inappropriate language

> *Inappropriate language:* As an employee of Fiscal for ten years, I am pretty fed up with its budget cuts and lay-offs. You administrative types are all alike—you do what you like with the company's money, and with Fiscal employees, and I just don't get your logic. So, you better make some changes, and make them fast.
>
> *Courteous language:* As an employee of Fiscal for ten years, I have witnessed several budget cuts like the one we are currently experiencing. While Fiscal has been experiencing some profit losses, I tend to wonder if budget cuts are the answer. Therefore, I'd like to know what alternatives existed (if any), and why these cuts became Fiscal's only option.

Both passages present relatively the same material. However, most readers would find the second passage more persuasive, since it requests a change in a more courteous manner, and it does not use inappropriate language, such as, "You administrative types." Even though the writer of the second passage had strong feelings about the topic, she maintained a professional tone.

Sexist language

Eliminating sexist language, which characterizes people solely on the basis of gender, is another important way to develop a courteous prose style. Eliminating sexist language in your prose helps you to maintain constructive working relationships with both male and female colleagues.

In some instances, properly identifying and avoiding sexist language has been mistakenly considered too difficult or confusing. Avoiding the use of sexist language is, on the contrary, relatively straightforward if you follow a few guidelines.

Avoid stereotypes, such as:

- All secretaries and nurses are women.
- All construction workers and CEOs are men.

Avoid masculine subject/pronoun overuse, such as:

- Using "Dear Sir:" in correspondence in which the addressee is unknown
- Using "he, his, him" to refer to humans in general—"Any successful professional knows that he must learn to communicate effectively."

Avoid title stereotyping, such as:

- Using "Chairman" rather than "Chairperson"
- Using "Mrs." rather than "Ms." in correspondence

Avoid patronizing labels, such as:

- lady doctor
- male nurse

Jargon

Jargon is language understood only by a specialized and select group. Eliminating jargon involves correctly identifying your audience by remaining aware of their level of expertise, education, and experience while you write the document.

For example, readers of the *Journal of the American Medical Association (JAMA)* are typically scientists and professionals within the medical field who understand and use the specialized, scientific language used in *JAMA*. However, a journalist or a lawyer—both trained professionals—would label this language as jargon. For them, the language used in *JAMA* is outside of their areas of expertise, and it is difficult, if not impossible, to understand its meaning. As a technical writer, you need to identify your readers' levels of expertise, and avoid using language that they may consider jargon.

Chapter 17 Task Exercise

Using *courteous* prose helps you to establish and maintain professional relationships with your clients, supervisors, and colleagues. To develop courteous prose, avoid biased or inappropriate language, sexist language, and jargon. Using what you have learned about these issues, revise the following passage for conciseness, clarity, *and* courtesy.

Once again, the following passage is an excerpt from the *Fiscal Workers Newsletter.* This excerpt is a "letter to the editor." If you were asked to revise it for clarity, conciseness, and courtesy, what changes would you make?

> ***Excerpt:*** A good Fiscal employee must be dedicated to the company, and he must work hard to produce a quality product. There are a lot of bad ideas going around the company about what to do to keep our employee morale up. With reference to those ideas, we need to simply stop being good employees. Those big bosses—every one of those guys—were not dedicated to the employees when they made those cuts, so we should not be dedicated to Fiscal, either.

To be readable and effective, your prose should be clear, concise, and courteous.

When your prose exhibits all of these qualities, the message of your document is better communicated to your audience. While the strategies for improving your prose style are challenging to implement, they'll become natural with practice, and you'll see your prose style improve.

Answer Key

Chapter 1 What Is Technical Communication?

Response Exercise 1.1

Responses to this exercise will vary, depending on the type of documents you choose to analyze. However, this exercise should show you the close connection among a company's products and services, the documents that describe or promote those products and services, and the company's reputation.

Response Exercise 1.2

Responses to this exercise will vary, depending on the type of documents you choose to examine. However, you should identify the type of reader toward which each document is geared. Also, you can begin to speculate about how those documents help to create or to sustain that company's reputation.

Response Exercise 1.3

Responses to this exercise will vary, depending on the type of document you choose to analyze. Whatever document you choose, you should be able to identify the five characteristics of technical communication in that document.

Chapter 1 Task Exercise

Responses to this exercise will vary, depending on the technical writer's responses. Below is an excerpt of the responses that resulted from an interview with a technical writer and a sample of that interview's corresponding Workplace Chart. Your own interview responses will differ from those presented in the sample; your responses probably will be more lengthy and descriptive.

Interview #1 Responses (excerpt)

The technical writer I interviewed, Rue Chan, has worked for nearly nine years at a federally funded agency. This agency manages the natural resources—animals, plants, and ecology—at a nature preserve in our area, and employs a number of administrative personnel, wildlife biologists, ecologists, and foresters.

While Rue writes many different types of documents, such as correspondence, brochures, and reports, she also presents information orally in formal and informal settings and maintains the agency's Web site. Rue produces *correspondence* for audiences in the organization, for other government agencies, and for the general public. She writes *brochures* not only to inform the public (who regularly visit the nature preserve) about the preserve's abundant natural resources, but also to instruct them about safety and environmental issues. Rue collaborates with her colleagues to write several different types of *reports*, usually to other government agencies, that may report on financial matters, environmental issues, or the effects of new federal policies on the preserve.

One type of document that Rue commonly creates is a natural resource informational brochure. Rue showed me one such brochure entitled, "Keeping Our Streams Clean," and she described the brochure in terms of the five characteristics of technical communication (see Sample Workplace Chart below).

Sample Workplace Chart

Document Type	*Communication Context*	*Content*	*Organization*	*Prose Style*	*Design*
Informational Brochure: "Keeping Our Streams Clean"	***Situation*** • Increase in tourism results in more stream pollution. • Need an inexpensive way to combat this problem. ***Purpose*** • Educate public about the beauty of our preserve's streams. • Help combat the pollution. ***Audience*** • Preserve visitors • Preserve volunteers	Main topics—geared to the layperson audience—answer the following questions: • What are *healthy* streams? • What types of elements pollute streams? • How quickly can streams *heal?* • How can you prevent pollution?	A variation of the cause/effect strategy. A discussion of the *causes* identifies the types of elements that pollute streams. A discussion of the *effects* reveals that not only streams, but also wildlife are affected.	Clear, concise, and courteous language. Stacked lists contain brief statements, facts, or questions threaded through each section of the brochure.	Three-panel, fold-out design. Full-color brochure printed on glossy paper. Contains maps and photos of streams and line graphs illustrating increase in pollution.

Chapter 2 What Do Technical Writers Do?

Response Exercise 2.1

1. Responses to this exercise will vary, depending on the type of information you gather. However, you should discover that nearly every industry employs technical writers.
2. For example, if you choose to describe a non-profit organization—one located at the county or federal level—you may realize that the organization is responsible not only for providing services to those who need them, but also for distributing information about those services or programs being offered. Technical writers, then, may help the organization communicate with those who are in need, and with those who volunteer their time or money to the organization.
3. Technical writers have career opportunities all over the world. (Note: Access the Society for Technical Communication [STC] Web site to locate Web sites describing overseas chapters.)
4. You may choose to work at nearly any type of company—from a computer-related or publishing agency to an engineering or non-profit firm. Employment with an international corporation, either in the U.S. or abroad, is also a possibility.

Response Exercise 2.2

1. Responses to this exercise will vary, depending on the types of colleagues you list. If you list a civil engineer, for example, that colleague performs a variety of duties related to the planning, design, construction, and maintenance of structures such as bridges, highway systems, and water supply and drainage systems. Civil engineers typically work with a number of different personnel—fellow engineers, project managers at construction sites, administrators, and clients.
2. A civil engineer must have a four-year degree.
3. Primarily, civil engineers use computers, not only to help them communicate, but also to problem-solve construction strategies and to design projects as well.
4. Communication—both written and oral—is important to civil engineers. They must be able to communicate their ideas and designs to a variety of audiences who have a variety of areas of expertise as well.
5. Civil engineers may help you to better decide what technical or design information to incorporate into a report or a specifications document.

Response Exercise 2.3

Responses to this exercise will vary, depending on the job listings you have obtained. However, you will probably discover that many different types of organizations require technical writers. You may find that computer skills for software such as MS Word, PowerPoint, Access, and Visio are required. Also, specific skills such as the ability to collaborate effectively, edit, or demonstrate knowledge in a related field also may be listed.

Chapter 2 Task Exercise

Responses to this exercise will vary, depending on the technical writer you choose and the responses he or she provides. Below is an excerpt from a letter to a career counselor that was created for this exercise.

Letter to a Career Counselor (excerpt)

Dear Ms. Smith:

The technical writer I interviewed, Rue Chan, has worked for nearly nine years at a federally funded agency. This agency manages the natural resources—animals, plants, and ecology—at a nature preserve in our area and employs a number of administrative personnel, wildlife biologists, ecologists, and foresters. As a technical writer, Rue has completed a four-year degree in professional communication; she has worked for this agency since she graduated.

Rue writes many different types of documents, such as correspondence, brochures, and reports. She produces correspondence for audiences within the organization, for other government agencies, and for the general public. She writes brochures not only to inform the public (who regularly visit the nature preserve) about the preserve's abundant natural resources, but also to instruct them about safety and environmental issues. Rue also presents information orally in formal and informal settings and maintains the agency's Web site.

Rue collaborates with her colleagues to write several different types of reports, usually to other government agencies, on financial matters, environmental issues, or the effects of new federal policies. As a technical writer, Rue may work with writers from other agencies; however, Rue collaborates with colleagues from her own agency a great deal. Rue works with personnel who have technical expertise, such as biologists, foresters, and ecologists, as well as with colleagues who have degrees in business—such as executive assistants and park administrators.

Each of Rue's writing projects is different; therefore, she works on each one in a slightly different way. One recent project, however—a revision of the park's nature trail guide—was particularly interesting to complete. The nature trail guide is a seven-page booklet illustrating the nearly two dozen nature trails accessible to visitors in the park. The guide also provides a complete list of park rules (regarding fires, litter, pets, etc.) and safety guidelines (including ways to deal with wildlife and basic hiking and first aid tips).

Rue began the project by closely reading the current nature trail guide. She then met with one of the park's ecologists, Jack, and a park administrator, Joanne. During this meeting, Rue and her colleagues identified and discussed ways to improve the nature trail guide. With several pages of revision suggestions, Rue began re-writing the guide and, after nearly a week, distributed a draft of the new guide to Jack and Joanne. Rue then received written feedback on the draft of the guide from both of the park employees.

Jack and Joanne were particularly pleased with the ways that Rue had revised the safety guidelines section of the guide: this revision was not only written more concisely, but also contained useful visual elements (headings, numbered lists, and illustrations) that made the text easy to scan. Both of Rue's colleagues, however, realized that several newly passed park rules should be included in the updated trail guide. Also, Joanne suggested that a visitor survey be added to the trail guide which would allow visitors to make suggestions for improving the trail system.

Rue spent the next several days implementing these suggestions, during which time she also asked Gail, Joanne's executive assistant, to read a draft of the trail guide and edit it for style and readability. After Rue incorporated Gail's suggestions, Rue again distributed the guide to Jack and Joanne. With approval from her colleagues, Rue was able to complete the revision of the nature trail guide.

Rue enjoys her job, and she had the following suggestions for those interested in a technical writing career. Rue notes that her interest in writing, communicating orally, and collaborating with colleagues, clients, and customers, helps her do her job more effectively. She advises that those interested in a technical writing career have similar interests as well. Also, Rue advises that technical writers should be willing to keep learning on-the-job and to continually develop their communication skills. Rue states that throughout their careers, those interested in technical writing should keep abreast of technological advances, attend professional conferences and seminars, and talk with other technical writers.

Chapter 3 What Do Technical Writers Produce?

Response Exercise 3.1

Responses to this exercise will vary, depending on the types of documents that you obtain. Based on the document's characteristics, you should be able to identify and describe it as either a type of report, a brochure, a letter, and so on.

Response Exercise 3.2

1. Responses to this exercise will vary, depending on the types of documents or resources you choose to analyze. If you list a Web site, for example, the *purpose* for the site may be to advertise "Company A's" products, while enabling customers to purchase those products directly. The *situation* that necessitated the site may have resulted from poor product sales or competitors' success on-line. The Web site's primary audience is potential and current customers.
2. The content of a Web site may contain useful product descriptions, helpful service information, or interesting company/personnel profiles.
3. For a Web site, you may want to sketch its *organization* by beginning with its home page and working through several of its major pages or links.
4. Web sites demand a clear, concise, and courteous prose style.
5. Appropriate design is important for both documents and Web sites. The Web site may be user-friendly in terms of its navigation (from one page to the next), its use of color, and its use of visual aids.

Chapter 3 Task Exercise

Responses to this exercise will vary, depending on the technical writer's responses. Below is an excerpt from a Genealogy Chart that was created for this exercise.

Sample Genealogy Chart

Informational Brochure:
"Keeping Our Streams Clean"

Situation
- Increase in tourism results in more stream pollution.
- Need an inexpensive way to combat this problem.

Purpose
- Educate public about the beauty of our preserve's streams.
- Help combat the excessive stream pollution.

Audience
- Preserve visitors.
- Preserve volunteers.

The brochure's beginning...	Rue and Jerry collaborate...	The brochure takes shape...
Through their research, both the preserve's wildlife biologists and ecologists find that streams in the preserve have become increasingly polluted. The main source of this pollution is the increase in tourist population. Jerry Halloran, a preserve wildlife biologist, obtains funding to produce a brochure. Jerry creates a rough outline of a brochure that will be geared for a preserve visitor audience, warning them of the effects of this type of pollution, while providing them with ways to stop the pollution. Jerry shows his outline to Rue.	Rue examines the outline Jerry created for the clean streams brochure. After Rue and Jerry discuss the main points that the outline presents, they conclude that some revisions need to occur. Rue suggests that the main points be worded in the form of questions. Also, Rue believes that an additional section should be included—one that discusses what a healthy stream should look like. Jerry agrees, but he notes that to "make room" for this additional section, the information in several other sections must be condensed.	Rue and Jerry discuss how to condense some of the information in the first two sections of the brochure. Jerry believes that all of this information is necessary, so Rue experiments with design options that may streamline this information without needing to eliminate very much of it. Susanne, the park administrator's executive assistant, shows them a similar type of brochure regarding trail safety. Rue decides to incorporate aspects of this brochure's design into the clean streams brochure.

Chapter 4 How Do You Become a Technical Writer?

Response Exercise 4.1

Responses to this exercise will vary, depending on your own interests, experiences, and attitudes toward education. You should create a complete and relatively lengthy (2–3 pages, handwritten) response to the four questions posed in this exercise. Oftentimes, prospective employers or a university admissions board will ask you to discuss—either orally or in writing—this type of information.

Response Exercise 4.2

Responses to this exercise will vary, depending on your own preferences and goals regarding time, financial investment, and program missions. Create complete and honest responses to each of these questions. Spend time responding to question 3, since knowing about each degree program's "mission" helps you to better understand how that program fits your needs.

Response Exercise 4.3

Responses to this exercise will vary, depending on the types of programs you choose to analyze. Choose programs in which you have a genuine interest. Identify how they seem to relate to those programs described in this section of the chapter—that is, do they offer a variety of degree program options, or do they offer less of a variety?

Chapter 4 Task Exercise

Responses to this exercise will vary, depending on the technical writer's responses. Below is an excerpt from a Letter of Reference that was created for this exercise.

Letter of Reference (excerpt)

Dear Mr. Lee:

Rue Chan has worked as a technical writer for nearly nine years at a federal agency that manages the natural resources—animals, plants, and ecology—of a nature preserve in our area. Rue's interests in understanding both communication and technology and preserving the environment prompted her to make this career choice. As a young girl, Rue traveled with her parents all over the world—from countries in Africa and Asia to parts of Australia and New Zealand. These early experiences of seeing exotic landscapes sparked Rue's early interest in the beauty and complexities of nature.

Rue completed a four-year degree in professional communication at New Mexico State University, and she recently enjoyed taking a Web design course, offered through a local community college.

Recently, Rue participated in a training seminar for other federal employees. The workshop, "Communicating with Colleagues," was geared toward enabling administrators to communicate more effectively with colleagues—particularly those colleagues whose areas of expertise are largely technically, rather than business oriented. Rue led a workshop that focused on better ways to lead department or agency meetings. As Rue noted, meetings such as these often involve a wide range of personnel, whose interests, goals, and backgrounds differ a great deal. Differences such as these often make communication and collaboration challenging.

Chapter 5 The What, Why, and Who of Technical Communication

Response Exercise 5.1

Responses to this exercise will vary, depending on your personal experiences. This exercise should show you that a variety of factors—including time and collaborators—often affect the documents you create.

Response Exercise 5.2

Responses to this exercise will vary, but the following responses are typical.

1. A video cassette recorder owner's manual included in each product package (a) instructs users about the product's operation and (b) informs them about a variety of accessories that they may purchase to enhance the product's use.
2. A letter from a local charity organization to its volunteers (a) reports about the organization's operation costs and the donations received through the last fiscal year and (b) persuades volunteers to pledge more of their time to the organization's causes.
3. A church newsletter circulated among church-goers (a) inquires about the interest in a Father's Day picnic and (b) recommends that the church-goers watch a television program about the effects of movie and video game violence on children.
4. A brochure for a local tourist attraction (a) persuades tourists not only to visit the city's botanical garden but also to attend its art museums and (b) informs them about the "Top Ten Best Restaurants" in the city, as rated by the local newspaper.

Chapter 5 Task Exercise

1. To understand this document's communication context, you should understand its purpose, audience, and the situation that necessitated it.

2. Questions to ask yourself about the situation that necessitated this manual's production include questions about time. For example, what's the deadline for this writing project? How many months, weeks, or days do I have to write this manual? How can I manage my time to meet this writing deadline?

   Then ask yourself questions about collaborators. For example, will I be working alone, or will I be writing within a team? If I am writing on a team, have I worked with members of this group before? What do I know about their work habits? What tasks is each group member responsible for completing? How will I communicate with my collaborators to produce a quality manual?

   Next, ask yourself questions about your experiences writing about a similar topic or writing a similar type of document. For example, what previous knowledge do I have about the topic of this manual? What knowledge do I need to begin drafting the document, and where and how will I access it? What previous knowledge—perhaps using other, similar processes—does my audience have about using this process? What are the needs of my audience—that is, as new employees learning a new process, what information do they need? Finally, ask yourself questions about the purpose of the manual. For example, what is its objective or goal? What work will this manual do, and why is it important?

Chapter 6 Organizing and Outlining Documents

Response Exercise 6.1

1. The users of an automobile instruction manual: A *chronological* organizing strategy is appropriate because it allows you to describe information, such as a series of events (how to best service your vehicle) or a set of briefer tasks (how to re-set the clock on your radio), as they occur or should occur in time.
2. Description of a new cordless telephone for a design department that must design suitable packaging for the new product: A *spatial* organizing strategy is appropriate for this task because it allows you to describe the cordless phone as it exists in space: top to bottom, side to side, or inside to outside.
3. Report describing the implementation of a new homeowner's insurance policy and discussing how the policy affects your company's overall policy sales: A *cause-and-effect* organizing strategy enables you to describe both the cause (implementation of new policy) and its effects on your company's sales.
4. Report recommending a strategy for improving customer service based on a number of factors you and your team have identified: An *order of importance* organizing strategy is useful for describing this strategy by first describing the factors your team has identified in order, from the most important to the least important.
5. Description of the new office paper recycling process, recently instituted, for those personnel who must oversee the recycling process within their own departments: A *divide-and-describe* organizing strategy is useful for describing a process, since it allows you to divide up the a process into a series of logical stages and describe each in turn.
6. Description of two new accounting software programs, and a recommendation of one that is best suited to your company's needs: A *compare/contrast* strategy emphasizes the similarities and differences of the two programs. Then you describe the program that is best suited to your company's needs.

Response Exercise 6.2

Responses to this exercise will vary, depending on your own experiences creating and working with outlines. Create complete and honest responses to each of these questions. Try to remember how you used an outline in the past and how you now plan to use one in the future.

Chapter 6 Task Exercise

1. To begin planning the report, you should remind yourself about the communication context. For example, what is the situation surrounding this document? Who are my collaborators? What deadlines do we have? Also, ask questions about the document's purpose and its audience. For example, what previous knowledge does my audience have about this topic? Then, create an outline of the major points that you need to convey. This outline could be a series of complete sentences, phrases, or key words that indicate each point that needs to be made.

2. The organizing strategies discussed in this chapter include cause and effect, chronological, compare/contrast, divide and describe, order of importance, and spatial.

   The strategy that is most useful for this recommendation report is a compare/contrast strategy. This strategy emphasizes the similarities and differences of the trucking company candidates, and makes the subsequent recommendation seem even more valid.
3. A possible outline for this recommendation report may include the following points:

Recommendation report outline

Introduction

A. Description of the project problem, tasks performed to solve this problem, and the purposes of this report.
B. A more in-depth discussion regarding the background of this project problem.
C. A more in-depth discussion about the tasks, including the research methods, used to research the three trucking companies.

Trucking Companies Description

D. A list and definition of all the criteria used to assess the trucking companies.
E. Description of Trucking Company A
F. Description of Trucking Company B
G. Description of Trucking Company C

Conclusion

H. Recommendation of best company
I. Summarize main points of the report

Chapter 7 Locating and Using Common Information Resources

Response Exercise 7.1

1. As a prospective vehicle buyer, you need more information about a particular vehicle's cost, performance, and efficiency.

   Outline

   A. Identify yourself as a prospective vehicle buyer, and indicate your purposes for writing this letter.
   B. Identify the vehicle (by model and year) about which you need more information, and create a stacked list of questions regarding that vehicle. Specifically, questions regarding its cost, performance, and efficiency.
   C. Include a response deadline, and conclude with a courteous close.
2. As a possible attendee, you need more information about the schedule, events, and purposes for a summer festival in your area.

   Outline

   A. Identify yourself as a possible summer festival attendee, and indicate your purpose for writing this letter.
   B. Identify the dates that you and your party plan to arrive at the festival. Then create a stacked list of questions about the upcoming festival regarding schedule, events, and purposes.
   C. Include a response deadline, and conclude with a courteous close.
3. As a likely customer, you need more information about the cost, schedule, and services provided by a home or lawn care service of a company in your area.

   Outline

   A. Identify yourself as a prospective lawn care service customer, and indicate your purposes for writing this letter.
   B. Describe your location, the size of your lawn, and any other necessary lawn or lawn care background. Then create a stacked list of questions regarding the lawn care services they provide, their cost, and possible schedules.
   C. Include a response deadline, and conclude with a courteous close.

Response Exercise 7.2

Responses to this exercise will vary. Here is a typical response for the hammer.

1. The claw and the handle. The claw is the heavy, metal object that is attached to the handle. The handle is a long, narrow, wooden handgrip for holding the hammer.
2. The claw contains two heavy, metal parts: the mallet and the forked end. The mallet is rounded, while the forked end juts from the mallet in two prongs.
3. The function of the mallet is to strike objects, while the function of the forked end is to pull out and remove nails. The claw is metal and roughly 4" from end to end, while the handle is wooden and roughly 12" long.

Chapter 7 Task Exercise

1. The six information resources that help writers locate more information include personal experience, interviews, request and response letters, surveys, objects or processes, and internal resources.
2. The following responses explain the (a) purpose of the resource, (b) situation where this resource is valuable, and (c) two strategies or tips that you should understand when using this resource.

   Your experience: (a) If you need more information about a topic, remembering and drawing upon your past experience is a useful place to start. (b) When writers need to refresh their memories about past documents or writing projects, they examine their archives (files of completed documents). These archives provide them with useful information about each document they have prepared. (c) Ask yourself these questions about your experience: Are any of my colleagues having similar or different experiences? How can I access information about these current experiences?

   Interviews: (a) An interview allows you to access the information you need directly from the source: the individual. (b) Interviews, when conducted properly, can be an efficient means to gain a wealth of information about a topic. (c) An effective interview, one that gives you enough of the right information, must pass through three stages: preparation, interview, and follow-up. After you have conducted the interview, send the individual a thank-you note.

   Request and response letters: (a) Another information resource that helps you to access the experience and knowledge of other people are letters of request and response. (b) Unlike an interview, correspondence allows you to get valuable information without actually meeting face-to-face with the individual. (c) Begin the request letter by identifying why the recipient is an important information source and how you will use the responses. Ask specific questions that are open ended but focused enough to generate productive responses.

   Surveys: (a) One of the best ways to access information from a large group of people is by sending them a survey to complete. (b) Whether there are five or five hundred respondents, a well-designed survey is a useful and inexpensive method for accessing information. (c) Ask yourself these questions about surveys: Do I have a target audience, or set of respondents, in mind for this survey? Have I asked an effective set of questions that are focused enough to generate specific responses yet open ended enough to allow for descriptive responses?

   Objects or processes: (a) One important way to understand as much as possible about the topic of a document is to locate and examine that topic yourself—particularly if that topic is an object or a process. (b) Many times, to best convince your audience about the value of an object or the importance of a process or set of tasks, it is necessary to describe thoroughly that object or process. (c) Obtain permission to examine, observe, photograph, sketch, or videotape the object or process. Ask questions about the object or process of those people who know the most about it.

   Internal resources: (a) Internal resources are information sites that can be found within your organization—a paper archive (several file cabinets or a whole room) or an electronic database. (b) Internal resources, whether stored electronically or in an archive, provide you with valuable, company specific information. (c) Your company's internal resource sites may contain information about the products or services that are provided by other companies. These sites may include information about your company's products or services as well.

Chapter 8 Incorporating Visual Aids Into Documents

Response Exercise 8.1

1. (a) Will the information be clearer and more understandable if I use a visual aid?
   (b) Will my inexperienced audience, who has never been exposed to this information before, learn more from a visual aid?
   (c) Will the interpretation of the information be more accurate if I use a visual aid?
   (d) Will the information I present in a visual aid complement, not stand in for, the discussion I am having?
2. First, technical writers strive to make their written text as clear and understandable as possible, and many times a visual representation of an object, process, or idea is a more effective way of communicating it.

   Second, using visual aids within a document may help those readers who are inexperienced with a concept or idea understand or learn about it more easily. Readers who do not know a great deal about the subject being discussed may learn more readily if a written discussion is accompanied by a visual aid.

   Third, visual aids can help readers understand and interpret information in a more accurate fashion. Unlike information formatted into a visual aid, some information is nearly impossible to scan quickly, since its paragraph organization does not help the reader accurately identify, compare, or understand the information.

   Fourth, visual aids are important parts of a document that may make reading and understanding the information that is being presented easier. However, you must understand that visual aids are almost always a complement to the written text and not a substitute for it.

Response Exercise 8.2

Responses to this exercise will vary, depending on the document and the visual aids contained in it. Responding thoroughly to each of these questions allows you to better understand the ways in which different visual aids are used for specific purposes within documents.

Chapter 8 Task Exercise

1. *Tables* are visual aids that organize data into columns and rows. Tables are most useful for displaying large amounts of numerical data. *Graphs* organize data in three ways: according to time (using a line graph), according to amount (using a bar graph), or according to percentage (using a pie graph). *Organizational charts* are useful visual aids that clearly represent the organization, or hierarchy, of a company, department, or project team. These types of charts normally use boxes and lines to depict organization. *Drawings* realistically portray objects and are most useful for showing detail or highlighting a certain point of view. *Photographs* are most useful to illustrate a more exact representation of an object. *Maps* show the geographical or "built" features of a region.
2. Sketch out each visual aid so you can identify its distinguishing features. Use the visual aids found in this chapter as examples.
3. Responses to this exercise will vary, but the following responses are typical. *Tables* organize numerical data, such as those financial data found in a corporation's annual report. *Pie graphs* organize data according to percentage, and they are useful for illustrating the different types of personnel who make up an organization. *Organizational charts* are useful for depicting the hierarchy of the personnel on a project team. *Drawings* realistically portray objects, such as a front view of a piece of equipment. *Photographs* represent objects, such as a new addition or wing to the company headquarters, accurately and realistically. *Maps* show the geographical or "built" features of a region, such as a map of a newly proposed waterfront development plan.

Chapter 9 Using Document Design Strategies to Create Usable and Appealing Documents

Response Exercise 9.1

Responses to this exercise will vary, depending on the type of user's manual that you choose to analyze. Responding thoroughly to each of these questions allows you to better understand the ways in which chunking and labeling work within a document.

Response Exercise 9.2

Responses to this exercise will vary, depending on the type of documents that you analyze. Responding thoroughly to each of these questions allows you to understand how the five spatial design elements may be used in different types of documents.

Response Exercise 9.3

Responses to this exercise will vary, depending on the type of documents that you analyze. Responding thoroughly to each of these questions allows you to understand how the textual design elements may be used in different types of documents.

Chapter 9 Task Exercise

Responses to this exercise will vary, depending on the type of article with which you decide to work. Create a complete sketch of the types of design elements that you would include to make the "op-ed" piece interesting and appealing. Also, responding thoroughly to each of the questions allows you to understand how both spatial and textual design elements may be used in a variety of documents.

Chapter 10 Collaborating With Colleagues and Writing Effectively Within a Group

Response Exercise 10.1

1. While collaborating on a document is more labor- and time-intensive than writing that document individually, you benefit from producing a collaboratively written document in several ways. First, spending quality time with your peers on an important project helps you to cultivate or strengthen your professional relationships with them. Second, observing how others approach the writing process and how they use written documents to solve communication problems helps you to learn more about yourself as a writer. Third, working with colleagues from a variety of departments helps you to learn more about different fields.
2. Your collaborators may work in a variety of areas within your organization: engineering, graphic design, publication, communication, or marketing. Once again, working with colleagues from a variety of fields helps you to learn more about those fields.
3. First, a document that is written using the combined efforts of a team of people who have different perspectives and a variety of talents and interests is a richer and more effective document than one produced by an individual. Second, a team of colleagues who collaborate on a document may also devote more energy and time to planning, drafting, and revising that document than would be possible for an individual writer.

Response Exercise 10.2

1. Two ways that you can prepare for your first team meeting include performing assigned tasks and creating a meeting agenda.
2. A list of possible discussion topics for your first planning meeting may include devising an outline of the brochure, delegating individual tasks, and planning your agenda for the next meeting.
3. To be an effective collaborator during this team meeting, you should show enthusiasm, be cooperative, listen, and ask questions.
4. To be an effective collaborator during this team meeting, you should reflect on what was said and done during the meeting, communicate with team members, and communicate with others, such as supervisors or clients, about your team's progress.

Response Exercise 10.3

1. *Procedural conflict* involves disagreements about factors including when to meet, how meetings should be run, and how decisions should be made. While this kind of conflict can be destructive, you can usually prevent it from happening by careful preliminary decisions. Procedural conflict issues such as this one should be dealt with openly and immediately by all members of the project team.

2. *Affective conflict* involves interpersonal disagreements that involve your personality, values, attitudes, and biases. This kind of conflict can be very destructive if you do not recognize it in yourself and in others.

   One way to combat affective conflict is to concentrate on making your team's product, the written document, as successful as possible. Rather than focusing on other team members and your attitudes towards them, try to focus on the document that your team is creating. Another way to combat affective conflict is to identify your own prejudices and biases, and then make a conscious decision to leave them at the door when you are working with collaborators.
3. *Substantive conflict* involves the decisions your team makes about the elements included in the document. Substantive conflict is helpful because it leads to consideration of a range of ideas and usually helps your team to create a better document.

   One of the best ways to make substantive conflict work for your team is to promote substantive conflict within your team's document planning sessions by asking critical questions about the document. Also, whenever necessary, play "devil's advocate" to help your team create a better document.
4. *Affective conflict* involves interpersonal disagreements that involve your personality, values, attitudes, and biases. In this instance, after the meeting, you should take that team member aside, and politely voice your discomfort and annoyance with the joke telling that occurred during the team meeting.
5. *Substantive conflict* is helpful because it leads to consideration of more ideas and usually helps your team to produce a better document. In this instance, accept your team member's critiques as a type of substantive conflict. Build on your team member's ideas by allowing your colleague to challenge your decisions about the document.

Chapter 10 Task Exercise

1. To prepare for the planning meeting, perform your assigned tasks, be prompt, and create a meeting agenda. To perform effectively during the meeting, show enthusiasm, be cooperative, listen, and ask questions. To plan for your next team meeting, reflect, communicate with your team, and communicate with others.
2. First, identify the team members who are present, clearly note the meeting's purpose, and date the meeting minutes. Second, identify and discuss each important point raised by the team. Describe this discussion completely enough so that people who did not attend the meeting will understand what occurred. Third, indicate the actions or tasks resulting from the discussion. Fourth, create a consistent organizational format for the meeting minutes so they are easy to scan and understand.

   You can communicate these meeting minutes to the group via e-mail or by circulating a memo among the members.
   (a) *Procedural conflict* involves disagreements about factors such as when to meet, how meetings should be run, who should be in charge (if anyone), and how decisions should be made. *Affective conflict* involves interpersonal disagreements that involve your personality, values, attitudes, and biases. *Substantive conflict* involves the decisions your team makes about the elements included in the document.
   (b) *Substantive conflict* is helpful because it leads to consideration of more ideas and usually helps your team to produce a better document. One of the best ways to make substantive conflict work for your team is for all members to be aware of the benefits of this type of conflict.
   (c) One member of your team begins to gossip about a fellow team member who is not present. You believe her gossip is false, but this discussion does not directly impact the brochure your team has been asked to create. What do you do?

Chapter 11 Writing Effectively by Understanding Planning and Drafting

Response Exercise 11.1

1. Writers may benefit from implementing time management strategies for a number of reasons. Mainly, though, technical writers must learn to produce quality documents in a short period of time; therefore, as a writer you must understand how to most effectively manage your time.
2. Three important questions you should ask yourself about time management include: (a) How much time do I have to produce a quality document? (b) What tasks are on my schedule now, and how will I prioritize these tasks with

these additional document planning and drafting tasks? (c) How long will it take me to plan and draft this document?

3. Three useful methods to help you manage and schedule your time include a simple day planner, scheduling software, or a Gantt chart. A Gantt chart lets you identify and label small, manageable tasks that are part of an overall project, assign each of these tasks a time frame and completion deadline, and input these tasks into the overall timeframe allowed for the project.
4. Responses will vary to this question; however, you should reflect on the ways you have (and have not) used time management strategies in the past.

Response Exercise 11.2

1. The three phases that make up invention and planning are (a) identifying the communication context, (b) brainstorming, and (c) organizing/designing. Two important questions you should ask yourself about each phase are listed below. (Note: You may choose to mention other types of questions than those listed.)
   (a) What is my document's situation? What is my document's purpose?
   (b) What information does my audience need and what information should I exclude? What is the argument I am making in this document?
   (c) What organization strategy is most suitable for ordering and discussing these points? What design strategy is suitable for my communication context?
2. Most writers move through these phases at some point during the writing process. While you should not think of these phases as a type of lock-step formula to pass through when creating a document, you should begin to ask yourself questions about each of these stages during the writing process.
3. Responses to this question will vary; however, you should try to reflect on the ways that you have passed through phases such as these while creating a document.

Chapter 11 Task Exercise

Responses to this exercise will vary, depending on the type of document you describe and your individual writing process. Your description of the process and your responses to the three questions should be honest and complete.

Chapter 12 Writing Effectively by Understanding Revising, Editing, and Document Cycling

Response Exercise 12.1

1. The four categories that you should check for during revision include (a) accurate and complete content, (b) attention to audience, (c) clear and logical organization, and (d) attention to mechanics. Two important questions that you should ask yourself about each phase are listed below. (Note: You may choose other types of questions than those listed.)
   (a) Are the facts that I am presenting throughout the document logical and accurate? Have I explained each major point completely?
   (b) Are the examples in my document appropriate for my audience? Is my tone appropriate for my audience?
   (c) How well does my organization frame and support my document's objective and its major points? What other strategy could be used to better organize the document or a section of the document?
   (d) Have I attended to spelling and usage? Have I attended to proper grammar and word choice?
2. You should revise for all four of these categories, since each category pertains to an important aspect of the document. Neglecting to revise for all four categories may mean the difference between producing an effective or ineffective document.
3. Responses to this question will vary. However, you should try to reflect on the ways that you have revised previous documents and the types of changes you can implement to make this stage even more productive.

Response Exercise 12.2

1. These are four questions that you should ask yourself during editing:
   (a) Are the message and content of the document accurate?
   (b) Are the message and content organized logically?
   (c) Is the document designed accurately and in a consistent fashion?
   (d) Do the mechanical errors detract from the professional tone?
2. Responses to this question will vary; however, you should try to reflect on the ways that you have edited previous documents and the types of changes you can implement to make this stage even more productive.

Chapter 12 Task Exercise

1. Document cycling enables a writer to receive suggestions from a variety of readers about specific ways to improve a document before that document is submitted to its primary audience. Document cycling is used in organizations as a way for writers to receive feedback about a document from a variety of different readers.
2. Document cycling is a process that is performed with a number of different reviewers, while revision is a process that an individual writer may engage in without any collaboration from colleagues.
3. Typically, the types of colleagues involved in document cycling include editors or other technical writers; personnel with specialized technical knowledge, such as engineers or designers; personnel who work within the budget or finance departments; lawyers or other legal personnel; or supervisors who have business or managerial degrees.
4. Document cycling improves the flow of ideas from one department to another, produces a rich and diverse set of commentary and suggestions for document improvement and change, and may lead to a re-evaluation of the efficacy of the organization's current communication network.

Chapter 13 Testing Documents for Accuracy and Usability

Response Exercise 13.1

While responses will vary, listed below are several possible audiences interested in each subject.

(a) The audience types interested in the environmental impact of the new paper mill may include homeowners living near the paper mill, paper mill administrators, other paper mill workers, and environmental activists.
(b) The audience types interested in the violence ratings for video games may include parents, video game users, store owners who sell video games, and teachers.
(c) The audience types interested in financial support for the local symphony may include current supporters of the local symphony, symphony members, and members of the local (city or county level) arts council.
(d) The audience types interested in the mistreatment of the elderly at an area nursing home may include concerned relatives of nursing home patients, the elderly, nursing home administrators, and nursing home workers.

Responses to the questions will vary. However, listed below are sample responses to subject "a" above.

1. Homeowners may be interested in the environmental impact—specifically, any health issues—regarding the paper mill. Paper mill administrators may have an interest in the economic or legal issues regarding the mill's environmental impact. Paper mill workers may have an interest in the economic issues—particularly the effects these issues have on their jobs. Environmental activists may be interested in the general environmental issues regarding that specific area or region of the country.
2. The homeowner's interest may be different from the mill administrator's interests in that the former may be concerned about health or quality of life issues, while the administrator may be more concerned with economic and legal issues impacting the mill. Each interest is similar in that both concerns are a result of the presence of the mill.

Response Exercise 13.2

Documents

(a) For an on-line tutorial for a computer software program, user-based testing could be performed.
(b) For an instruction manual to accompany a combination coffee bean grinder/coffee maker, user- and text-based testing could be performed.
(c) For an informational brochure describing the policies and procedures of a new home-to-clinic shuttle service available for patients with limited transportation options, text- and expert-based testing could be performed.
(d) For a Web site that describes and promotes a county-run recreation vehicle campground, user- and text-based testing could be performed.

Questions

1. *Text-based* usability testing assesses the textual and visual features of a document—literally the words, sentences, and visual aids that make up the document.

   Expert-based testing involves the participation of one or more levels of review—from formal to informal readers. Technical readers assess the document for technical and content accuracy. Style readers assess a document's overall communicative accuracy.

   User-based usability testing involves watching and listening while a document user reads aloud and uses the document, or it involves asking the user questions about the document's usability after he or she has read and used the document.
2. *Text-based* testing is useful for evaluating how effectively and efficiently a reader can read the text. *Expert-based* testing is useful for assessing a document's technical accuracy, its stylistic and audience readability, and its legal soundness. *User-based* testing is useful for assessing the accuracy and usability of a variety of different types of documents.

To persuade your supervisor to fund usability testing, you may try conducting a usability test to pique your colleagues' interest. Often observing how and why a document fails or succeeds is an eye-opening and interest-generating process.

Chapter 13 Task Exercise

1. The usability testing ideally performed on an on-line instruction manual for a new software program is user-based testing. This type of testing involves watching and listening while a user reads aloud and uses the document, or it involves asking the user questions about the document's usability after he or she has read and used the document.
2. Including usability testing in this writing process is beneficial to your company's reputation, its product costs, and personnel costs. Company reputation is positively impacted, since satisfied users tend to spread their product satisfaction by word-of-mouth, thus strengthening your company's reputation. Product costs, such as the costs of a software application that your employees use, can be lessened by decreasing reparative costs, such as software updates or maintenance releases. Personnel costs, particularly those involved in maintaining a customer service line or training new employees, can be lessened by a rigorous usability testing program.
3. The steps of a typical user-based testing procedure involve selecting subjects, performing the test, and providing feedback to the document writer. Select a subject who has never read through the document before. Then ask him or her to read the document out loud and to talk through the decisions being made (as a result of reading the document). Remember to ask the user to comment on the document's clarity and effectiveness during the user testing process itself.

Chapter 14 Working With a Designer and Printer to Produce Documents

Response Exercise 14.1

Responses to this exercise will vary, depending on the type of design elements that you choose to include in the newsletter. Responding thoroughly to each of the questions allows you to understand the effects of these design elements—and the use of publication software—on your document.

Response Exercise 14.2

Responses to this exercise will vary, depending on the type of articles you choose to analyze. Responding thoroughly to each of the questions allows you to understand how design features are used in different types of publications.

Response Exercise 14.3

Responses to this exercise will vary, depending on the type of articles that you choose to analyze. Responding thoroughly to each of the questions allows you to understand how features such as choice of paper, color, layout, and binding affect the "look" of documents.

Chapter 14 Task Exercise

Responses to this exercise will vary, depending on the printing firm that you visit and the printer you ask. However, the following responses are typical of those you will probably receive.

1. The printing firm may advertise its services to prospective customers in a number of different ways. Typically, printers place advertisements in local newspapers and magazines and perhaps on television or the radio. Also, an attractive and usable Web site may help the firm sell its services. Like many businesses, printing firms also rely on word-of-mouth to generate sales.
2. A wide range of topics may be discussed with a prospective customer, and the topics discussed reflect what type of printing work is being requested. For example, printers may discuss basic services, costs, document turnaround, and preferred customer incentives. Also, printers may discuss and show you past printing projects that are similar to the work that your company may request.
3. Typically, the printer and a first-time customer would need to discuss a number of different topics to print 150 copies of a brochure. The printer and customer must discuss the kind of paper to be used, the use and type of ink color, page layout and design, and distribution methods. They must also agree on such issues as time constraints and cost.

Chapter 15 Understanding Expectation and Interpretation in Reading

Response Exercise 15.1

1. Here are three elements that you should understand about the ways in which your audience reads:
   (a) Readers rely heavily on prior knowledge—what they already know about a subject—to guide and inform their understanding of the document.
   (b) Readers demand texts that make sense to them—that is, texts that are purposeful, well-organized, and present content in a logical manner.
   (c) Readers read for certain purposes, whether they consciously realize it or not.
2. A reader usually has multiple purposes for reading a single document. Therefore, a reader may skim, read to learn, and read to do.
3. Readers *skim* a document to decide whether they want or need to read it more thoroughly. Also, readers determine the parts of the text that require concentration, that can be skipped, or that should be read first. Readers who *read to learn* are searching to understand meaningful information, concepts, and ideas. Readers who *read to do* use the information they have read or are reading to perform a task or process.

Response Exercise 15.2

1. The four strategies that you can use to read more effectively include manipulating your environment, recognizing genres, pausing to question, and remembering to reflect.
   (a) Your reading environment should be conducive to reading. One way to make it more effective is to identify the distractions in a particular area, eliminate the distractions or move to another area, and read exclusively in that suitable environment.
   (b) Knowing the genre means that you use what you know about the genre—the type of document—to help shape your reading and understanding of that document.
   (c) Readers stop periodically during the reading process and question themselves about the material they have just read.

(d) Readers reflect on the material they have read not simply to understand it, but also to interpret, analyze, and critique this information. Readers use several different methods to help them reflect—such as careful note-taking, synthesis, and analysis.

2. (a) Manipulating your reading environment enables you to pay close attention to the details and ideas presented in the document.
   (b) Knowing the genre allows you to use what you know about the genre to help shape your reading and understanding of that document.
   (c) Readers stop periodically during the reading process to question themselves about the material they have just read in order to ensure that they fully understand the concepts and ideas being introduced.
   (d) Readers may retain more information more accurately by reflecting several times for each document.
3. The three processes you can follow to help you better reflect include careful note-taking, synthesis, and analysis.

Responses to the final part of the exercise (questions 4–6) will vary, depending on the type of document that you choose to read and analyze. Responding thoroughly to each of the questions allows you to understand how you read, and how you may need to change your reading habits.

Chapter 15 Task Exercise

Responses to this exercise will vary, depending on the type of document that you choose to analyze. Responding thoroughly to each of the questions allows you not only to understand how you read, but also how reading affects the choices a writer should make while writing a document.

Chapter 16 Creating Logical and Coherent Documents

Response Exercise 16.1

1. In order to choose an organization strategy that will allow you to create a document that is logical and coherent, you need to understand that document's communication context. After identifying this context, you should choose a strategy that is suitable to your audience, the situation, and the document's purpose.

   An organization strategy dictates the order of points in a document, and to some degree, the emphasis placed on each of these points. Therefore, the logic and coherence that a document does or does not exhibit can be attributed largely to its organization.
2. To find out information about a document's audience, ask yourself the following questions:
   (a) What is the type of audience: clients, customers, co-workers, supervisors?
   (b) What previous knowledge does my audience have about this topic?
   (c) What work do they do?
   (d) What are their needs and interests in reading this document?
   (e) What does my audience value? What are their biases?
   (f) When, where, and how will my audience use this document?
3. The *cause-and-effect* strategy enables you to describe both the cause and effect(s) of an event, idea, or recommendation. A *chronological* organizing strategy allows you to describe information, such as a series of events or a set of tasks, as they occur or should occur in time. This strategy is most often used when technical writers draft instruction manuals or write documents that describe a process. The *compare/contrast* strategy emphasizes the similarities and differences of the subjects of a document, whether they are particular topics, objects, or events.

Response Exercise 16.2

Responses to this exercise will vary, depending on the type of document that you choose to analyze. Responding thoroughly to each of the questions allows you to understand how visual cues work in documents.

Chapter 16 Task Exercise

Responses to this exercise will vary, depending on the type of document you choose to analyze. Responding thoroughly to each of the questions allows you to understand how visual and textual cues work in documents.

Chapter 17 Writing Readable Sentences

Response Exercise 17.1

Excerpt: The recent budget cuts at Fiscal have generated a lot of talk lately. Several people in our town—many of whom work for Fiscal—sense these changes. Through discussions with many of them, we've found that the budget cuts have impacted their lives in three important ways: a sense of hopelessness about the future, a feeling of betrayal, and anger.

Response Exercise 17.2

Excerpt: The recent budget cuts may be another example of one of Fiscal's economic restructuring processes. Perhaps Fiscal has more ideas about future restructuring. Since future cuts or layoffs are possible, making any personal, economic plans for the future is difficult, as a Fiscal employee. My future economic plans have been cancelled.

Chapter 17 Task Exercise

Excerpt: Good Fiscal employees must be dedicated to the company; they must work hard to produce a quality product. Poor ideas regarding the improvement of employee morale continue to circulate throughout the company. To truly improve our morale, we should stop being model Fiscal employees. During the budget cut process, Fiscal's top-level administrators were not dedicated to other, subordinate employees. Therefore, we subordinate Fiscal employees should not be dedicated to Fiscal, either.

Appendix A

Technical Communication Education Programs

Graduate Level Programs

The following list provides you with a sample of the graduate level degree programs available in the United States. While this list is not an exhaustive one, it is a good place to begin your program search. Most graduate level degree programs also offer four-year programs in technical communication, so this list is a good place to start searching for these types of programs as well.

Each degree program entry includes the program's contact information, including mailing address, phone number, and e-mail address (when available), and a Web site address for the institution. (Note: Institution Web site addresses current for summer 1999. All contact information compiled from *Peterson's Graduate Programs in the Humanities, Arts, and Social Sciences*, 1999.)

Arkansas at Little Rock, University of
College of Arts, Humanities, and Social Sciences
Department of Rhetoric and Writing
Little Rock, AR 72204-1099
Application Contact: Dr. Charles Anderson, Coordinator
Phone: (501) 569-3160
URL: http://www.ualr.edu/

Boise State University
College of Arts and Sciences
Department of English
Program in Technical Communication
Boise, ID 83725-0399
Application Contact: Dr. Mike Markel, Director
Phone: (208) 385-3088
URL: http://www.idbsu.edu/

Bowling Green State University
College of Arts and Sciences
Department of English
Program in Scientific and Technical Communication
Bowling Green, OH 43403
Application Contact: Dr. Alice Philbin, Director
Phone: (419) 372-7552
URL: http://www.bgsu.edu/

Carnegie Mellon University
College of Humanities and Social Sciences
Department of English
Program in Professional Writing
Pittsburgh, PA 15213-3891
Application Contact: Paul Hopper, Director of Graduate Studies
Phone: (412) 268-2850
E-mail: hopper@andrew.cmu.edu
URL: http://www.cmu.edu/

Central Florida, University of
College of Arts and Sciences
Program in English
Orlando, FL 32816
Application Contact: Dr. Gerald Schiffhorst, Coordinator
Phone: (407) 823-2276
URL: http://www.ucf.edu/

Clemson University
Department of English
801 Strode Tower
Box 341503
Clemson, SC 29634-1503
Application Contact: Graduate Program Director
Phone: (864) 656-3151
URL: http://www.clemson.edu/

Colorado State University
College of Liberal Arts
Department of Journalism and Technical Communication
Fort Collins, CO 80523-0015
Application Contact: Donna Rouner, Coordinator
Phone: (970) 491-5556
E-mail: drouner@vines.colostate.edu
URL: http://www.colostate.edu/

Colorado at Denver, University of
College of Liberal Arts and Sciences
Program in Technical Communication
Denver, CO 80217-3364
Application Contact: Renee Combs, Administrative Assistant
Phone: (303) 556-8304
URL: http://www.cudenver.edu/

Drexel University
College of Arts and Sciences
Program in Technical and Science Communication
3141 Chestnut Street
Philadelphia, PA 19104-2875
Application Contact: Program in Technical and Science Communication
Phone: (214) 895-1823
URL: http://www.drexel.edu/

Eastern Michigan University
College of Arts and Sciences
Department of English Language and Literature
Programs in English
Ypsilanti, MI 48197
Application Contact: Dr. Elizabeth Daumer, Coordinator
Phone: (734) 487-4220
URL: http://www.emich.edu/

Florida Institute of Technology
College of Science and Liberal Arts
Department of Humanities
Program in Technical and Professional Communication
Melbourne, FL 32901-6975
Application Contact: Carolyn P. Farrior, Associate Dean of Graduate Admissions
Phone: (407) 674-7118
E-mail: cfarrior@fit.edu
URL: http://www.fit.edu/

Illinois Institute of Technology
Armour College of Engineering and Sciences
Department of Humanities
Chicago, IL 60616-3793
Application Contact: Graduate College
Phone: (312) 567-3024
E-mail: grad@minna.cas.iit.edu
URL: http://www.iit.edu/

Iowa State University
College of Liberal Arts and Sciences
Department of English
Ames, IA 50011
Application Contact: Dr. Kathleen Hickok, Director of Graduate Education
E-mail: englgrad@iastate.edu
Phone: (515) 294-2477
URL: http://www.iastate.edu/

James Madison University
College of Arts and Letters
Program in Technical and Scientific Communication
Harrisonburg, VA 22807
Application Contact: Dr. Peter J. Hager, Director
Phone: (540) 568-8018
E-mail: hagerpj@jmu.edu
URL: http://www.jmu.edu/

Mercer University, Cecil B. Day Campus
Graduate Engineering Programs
3001 Mercer University Drive
Atlanta, GA 30341-4155
Application Contact: Dr. David Leonard, Director of Admissions
Phone: (770) 986-3203
URL: http://www.mercer.edu/cbd/

Miami University (Ohio)
College of Arts and Sciences
Department of English
Program in Technical and Scientific Communication
Oxford, OH 45056
Application Contact: Dr. Robert R. Johnson, Director of Graduate Study
Phone: (513) 529-7530
URL: http://www.muohio.edu/

Michigan Technological University
College of Sciences and Arts
Department of Humanities
Houghton, MI 49931-1295
Application Contact: Dr. Diane Shoos, Director
Phone: (906) 487-3248
E-mail: dshoos@mtu.edu
URL: http://www.mtu.edu/

Michigan, University of
College of Engineering
Program in Technical Information Design and Management
Ann Arbor, MI 48109
Application Contact: Program Office
Phone: (734) 764-1426
URL: http://www.umich.edu/

Minnesota Twin Cities Campus, University of
College of Agricultural, Food, and Environmental Sciences
Department of Rhetoric
St. Paul, MN 55108
Application Contact: Dr. Ann Hill Duin, Director of Graduate Studies
Phone: (612) 624-4761
E-mail: rhetoric@tc.umn.edu
URL: http://www.umn.edu/

Montana Tech of The University of Montana
Graduate School
College of Humanities, Social Sciences, and Information Technology
Butte, MT 59701-8997
Application Contact: Cindy Dunstan, Administrative Assistant
Phone: (406) 496-4128
E-mail: cdunstan@po1.mtech.edu
URL: http://www.mtech.edu/

New Jersey Institute of Technology
Department of Humanities and Social Sciences
Program in Professional and Technical Communication
Newark, NJ 07102-1982
Application Contact: Kathy Kelly, Director of Admissions
Phone: (973) 596-3300
E-mail: admissions@njit.edu
URL: http://www.njit.edu/

New Mexico State University
College of Arts and Sciences
Department of English
Las Cruces, NM 88003-8001
Application Contact: Department of English
Phone: (505) 646-3931
URL: http://www.nmsu.edu/

North Carolina State University
College of Humanities and Social Sciences
Department of English
Raleigh, NC 27695
Application Contact: Dr. Robert V. Young, Director of Graduate Programs
Phone: (919) 515-4107
URL: http://www.ncsu.edu/

Northeastern University
Graduate School of Arts and Sciences
Department of English
Program in Technical and Professional Writing
Boston, MA 02115-5096
Application Contact: Kathryn Goodfellow, Assistant to Graduate Programs
Phone: (617) 373-3692
E-mail: kgoodfel@lynx.dac.neu.edu
URL: http://www.neu.edu/

Oklahoma State University
College of Arts and Sciences
Department of English
Stillwater, OK 74078
Application Contact: Department of English
Phone: (405) 744-9473
URL: http://www.okstate.edu/

Oregon State University
Graduate School
College of Liberal Arts
Program in Scientific and Technical Communication
Corvallis, OR 97331
Application Contact: Dr. Simon S. Johnson, Director
Phone: (541) 737-1650
URL: http://www.orst.edu/

Polytechnic University, Brooklyn Campus
Department of Humanities and Social Sciences
Major in Specialized Journalism
Six Metrotech Center
Brooklyn, NY 11201-2990
Application Contact: John S. Kerge, Dean of Admissions
Phone: (718) 260-3200
E-mail: admitme@poly.edu
URL: http://www.poly.edu/

Purdue University
School of Liberal Arts
Department of English
West Lafayette, IN 47907
Application Contact: Dr. A.W. Astell, Director, Graduate Studies
Phone: (765) 494-3748
URL: http://www.purdue.edu/

Rensselaer Polytechnic Institute
School of Humanities and Social Sciences
Department of Language, Literature, and Communication
Program in Technical Communication
Troy, NY 12180-3590
Application Contact: Kathy A. Colman, Recruitment Coordinator
Phone: (518) 276-6469
E-mail: colmak@rpi.edu
URL: http://www.rpi.edu/

San Jose State University
College of Humanities and Arts
Department of English
San Jose, CA 95192-0001
Application Contact: Dr. Don Keesey, Graduate Advisor
Phone: (408) 924-4435
URL: http://www.sjsu.edu/

Southern Polytechnic State University
Program in Technical and Professional Communication
South Marietta Parkway
Marietta, GA 30060-2896
Application Contact: Humanities and Technical Communication Department
Phone: (770) 528-7202
URL: http://www.spsu.edu/

Texas Tech University
Graduate School
College of Arts and Sciences
Department of English
Lubbock, TX 79409
Application Contact: Department of English
Phone: (806) 742-2501
URL: http://www.ttu.edu/

Utah State University
College of Humanities
Arts and Social Sciences
Department of English
Logan, UT 84322
Application Contact: Dr. Kenneth W. Brewer, Director of Graduate Studies
Phone: (435) 797-2733
URL: http://www.usu.edu/

Washington, University of
College of Engineering
Department of Technical Communication
Seattle, WA 98195
Application Contact: Anita Smith, Administrator
Phone: (206) 685-1558
URL: http://www.washington.edu/

Appendix B

Additional Resources

Professional Societies

The following list provides you with a sample of the professional societies to which technical writers may belong. While this list is not an exhaustive one, it gives you a good idea of the types of active professional societies that technical writers may join.

Joining a professional society is beneficial for a number of different reasons. For example, societies such as these provide you with education and training information, schedules of conferences and seminars, valuable business contacts, and a forum in which you can discuss relevant professional issues.

Each entry includes contact information, including mailing address, phone number, and e-mail address (when available), and a Web site address for the society. (Note: Institution Web site addresses current for summer 1999. All contact information compiled from *The St. Martin's Bibliography of Business and Technical Communication*, 1997.)

American Society for Training and Development
Box 1443
1640 King Street
Alexandria, VA 22313
Phone: (703) 683-8100
E-mail: info@ASTD.org
URL: http://www.astd.org/

Association for Business Communication
Baruch College, CUNY
Department of Speech
Box G1326
17 Lexington Avenue
New York, NY 10010
Phone: (212) 387-1340
E-mail: info@theabc.org
URL: http://www.theabc.org/

The Association of Computing Machinery Special Interest Group for DOCumentation (SIGDOC)
ACM Member Service Department, Association for Computing Machinery
P.O. Box 12115
Church Street Station, NY 10249
Phone: (212) 626-0500
E-mail: acmhelp@acm.org
URL: http://www.acm.org/sigdoc/

Association of Teachers of Technical Writing
Billie J. Wahlstrom, ATTW
Department of Rhetoric
201 Haecker Hall
1364 Eckles Avenue
University of Minnesota
St. Paul, MN 55108-6122
URL: http://english.ttu.edu/ATTW/

Council for Programs in Technical and Scientific Communication
Karen R. Schnakenberg
Department of English
Carnegie Mellon University
5000 Forbes Avenue
Pittsburgh, PA 15213
Phone: (412) 268-2659
E-mail: krs@andrew.cmu.edu

IEEE Professional Communication Society
Membership, IEEE Service Center
P.O. Box 1331
Piscataway, NJ 08555
Phone: (800) 678-IEEE
E-mail: info@ASTD.ord
URL: http://www.ieee.org/pcs/

Society for Technical Communication
901 North Stuart Street, Suite 904
Arlington, VA 22203-1854
Phone: (703) 522-4114
E-mail: stc@tmn.com
URL: http://www.stc.org/

Professional Journals

The following section lists those professional journals to which technical writers may subscribe. While this list is not an exhaustive one, it gives you a good idea of the types of professional journals of interest to technical writers.

Each entry includes the journal's contact address for subscription information and a brief journal description. (Note: All contact information compiled from journal Web sites and *The St. Martin's Bibliography of Business and Technical Communication*, 1997.)

Administrative Science Quarterly
ASQ
Cornell University
20 Thornwood Drive, Suite 100
Ithaca, NY 14850

Published four times per year. Includes articles and reviews of books focusing on administrative communication.

Business Communication Quarterly
Association for Business Communication
Department of Speech, Baruch College—CUNY
17 Lexington Avenue
New York, NY 10010

Published four times per year. Includes articles and reviews of books focusing on business communication education.

IEEE Transactions on Professional Communication
IEEE
345 East 47th Street
New York, NY 10017-2394

Published four times per year. Includes articles and reviews of books focusing on all forms of oral and written communication in the technological and scientific fields.

Information Design Journal
IDJ Subscriptions
P.O. Box 2230
Reading, RG5 4FH, England

Published three times per year. Includes articles and reviews that focus on print and on-line design issues.

Journal of Business Communication
Association for Business Communication
Department of Speech
Baruch College—CUNY
17 Lexington Avenue
New York, NY 10010

Published four times per year. Includes articles that focus on issues pertaining to business and organizational communication.

Journal of Business and Technical Communication
Sage Publications, Inc.
2455 Teller Road
Thousand Oaks, CA 91320

Published four times per year. Includes articles and reviews of books focusing on issues pertaining to business, technical, and scientific communication.

Journal of Technical Writing and Communication
Baywood Publishing Company
26 Austin Avenue
P.O. Box 337
Amityville, NY 11701

Published four times per year. Includes articles and reviews of books focusing on issues pertaining to business, technical, and scientific communication.

Science Communication
Sage Publications, Inc.
2455 Teller Road
Thousand Oaks, CA 91320

Published four times per year. Includes articles that examine issues relating to communication in scientific and technological fields.

Technical Communication
Society for Technical Communication
901 North Stuart Street
Suite 904
Arlington, VA 22203

Published four times per year. Includes articles that examine issues relating to professional communication in business, scientific, and technological fields.

Technical Communication Quarterly
Billie J. Wahlstrom, ATTW
Department of Rhetoric
202 Haecker Hall
1364 Eckles Avenue
University of Minnesota
St. Paul, MN 55108-6122

Published four times per year. Includes articles that examine issues relating to teaching professional communication.

Web Sites

The following section lists Web sites of interest to technical writers. While this list is not an exhaustive one, it gives you several professional resources that technical writers find useful.

American English Dictionary
gopher://gopher.niaid.nih.gov/77/deskref/.Dictionary/enquire

Hypertext *Webster* Interface
http://c.gp.cs.cmu.edu:5103/prog/webster

Roget's *Thesaurus*
http://humanities.uchicago.edu/forms_unrest/ROGET.html

William Strunk's *The Elements of Style*
http://www.columbia.edu/acis/bartleby/strunk/

Writers' Reference
http://longman.awl.com/englishpages/

http://www.writers-free-reference.com

Bibliography

Alred, Gerald J. (ed.). *The St. Martin's Bibliography of Business and Technical Communication.* NY: St. Martin's Press, 1997.

Burnett, Rebecca E. *Technical Communication*, 4th ed. Belmont, CA: Wadsworth, 1997.

Dumas, Joseph S., and Janice C. Redish. *A Practical Guide to Usability Testing.* Norwood, NJ: Ablex, 1993.

Houp, Kenneth W., Thomas E. Pearsall, and Elizabeth Tebeaux. *Reporting Technical Information,* 8th ed. Boston: Allyn and Bacon, 1995.

Nelson, Roy Paul. *Publication Design*, 5th ed. Dubuque, IA: Wm. C. Brown, 1991.

Peterson's Graduate Programs in the Humanities, Arts, and Social Sciences. Princeton, NJ: Peterson's, 1999.

Redish, Janice C. "Reading to Learn to Do." *The Technical Writing Teacher*, 1998. 15(3): 223–233.

Schriver, Karen A. "Evaluating Text Quality: The Continuum From Text-focused to Reader-focused Methods." *IEEE Transactions on Professional Communication*, 1989. 32(4): 238–255.

Society for Technical Communication. *1998 Technical Communicator Salary Survey.*

—*Careers in Technical Communication.*

—"Technical Writing One of 20 Hot Job Tracks." *News Release*, October 1998.